Exterior Differential Systems and Euler-Lagrange
Partial Differential Equations

Chicago Lectures in Mathematics Series

Editors: Spencer J. Bloch, Benson Farb, Norman R. Lebovitz, Carlos Kenig, and J. P. May

Other *Chicago Lectures in Mathematics* titles available
from the University of Chicago Press

Exterior Differential Systems and Euler-Lagrange Partial Differential Equations

Robert Bryant, Phillip Griffiths, and Daniel Grossman

THE UNIVERSITY OF CHICAGO PRESS · CHICAGO AND LONDON

Robert Bryant is the J. M. Kreps Professor in the Department of Mathematics at Duke University. **Phillip Griffiths** is director of the Institute for Advanced Study and professor in the Department of Mathematics at Duke University. **Daniel Grossman** was an L. E. Dickson Instructor in the Department of Mathematics at the University of Chicago at the time of writing, and is now a consultant at the Chicago office of the Boston Consulting Group.

The University of Chicago Press, Chicago 60637
The University of Chicago Press, Ltd., London
© 2003 by The University of Chicago
All rights reserved. Published 2003
Printed in the United States of America

12   11   10   09   08   07   06   05   04   03   1   2   3   4   5

Library of Congress Cataloging-in-Publication Data
Bryant, Robert L.
  Exterior differential systems and Euler-Lagrange partial differential equations / Robert Bryant, Phillip Griffiths, and Daniel Grossman.
    p. cm.—(Chicago lectures in mathematics series)
  Includes bibliographical references and index.
  ISBN 0-226-07793-4 (acid-free paper) —ISBN 0-226-07794-2 (pbk.: acid-free paper)
  1. Exterior differential systems. 2. Lagrange equations. I. Griffiths, Phillip, 1938–
  II. Grossman, Daniel Andrew, 1974– III. Title. IV. Chicago lectures in mathematics.
  QA649.B744 2003
  516.3'6—dc21   2003044775

Contents

Preface

During the 1996-97 academic year, Phillip Griffiths and Robert Bryant conducted a seminar at the Institute for Advanced Study in Princeton, NJ, outlining their recent work (with Lucas Hsu) on a geometric approach to the calculus of variations in several variables. The present work is an outgrowth of that project; it includes all of the material presented in the seminar, with numerous additional details and a few extra topics of interest.

The material can be viewed as a chapter in the ongoing development of a theory of the geometry of differential equations. The relative importance among PDEs of second-order Euler-Lagrange equations suggests that their geometry should be particularly rich, as does the geometric character of their conservation laws, which we discuss at length.

A second purpose for the present work is to give an exposition of certain aspects of the theory of exterior differential systems, which provides the language and the techniques for the entire study. Special emphasis is placed on the method of equivalence, which plays a central role in uncovering geometric properties of differential equations. The Euler-Lagrange PDEs of the calculus of variations have turned out to provide excellent illustrations of the general theory.

Introduction

In the classical calculus of variations, one studies functionals of the form

$$\mathcal{F}_L(z) = \int_\Omega L(x, z, \nabla z)\, dx, \qquad \Omega \subset \mathbf{R}^n, \tag{1}$$

where $x = (x^1, \ldots, x^n)$, $dx = dx^1 \wedge \cdots \wedge dx^n$, $z = z(x) \in C^1(\bar\Omega)$ (for example), and the *Lagrangian* $L = L(x, z, p)$ is a smooth function of x, z, and $p = (p_1, \ldots, p_n)$. Examples frequently encountered in physical field theories are Lagrangians of the form

$$L = \tfrac{1}{2}\|p\|^2 + F(z),$$

usually interpreted as a kind of energy. The *Euler-Lagrange equation* describing functions $z(x)$ that are stationary for such a functional is the second-order partial differential equation

$$\Delta z(x) = F'(z(x)).$$

For another example, we may identify a function $z(x)$ with its graph $N \subset \mathbf{R}^{n+1}$, and take the Lagrangian

$$L = \sqrt{1 + \|p\|^2},$$

whose associated functional $\mathcal{F}_L(z)$ equals the area of the graph, regarded as a hypersurface in Euclidean space. The Euler-Lagrange equation describing functions $z(x)$ stationary for this functional is $H = 0$, where H is the mean curvature of the graph N.

To study these Lagrangians and Euler-Lagrange equations geometrically, one has to choose a class of admissible coordinate changes, and there are four natural candidates. In increasing order of generality, they are:

- Classical transformations, of the form $x' = x'(x)$, $z' = z'(z)$; in this situation, we think of (x, z, p) as coordinates on the space $J^1(\mathbf{R}^n, \mathbf{R})$ of 1-jets of maps $\mathbf{R}^n \to \mathbf{R}$.[1]

- Gauge transformations, of the form $x' = x'(x)$, $z' = z'(x, z)$; here, we think of (x, z, p) as coordinates on the space of 1-jets of sections of a bundle $\mathbf{R}^{n+1} \to \mathbf{R}^n$, where $x = (x^1, \ldots, x^n)$ are coordinates on the base $\mathbf{R}^n$ and $z \in \mathbf{R}$ is a fiber coordinate.

---

[1] A *1-jet* is an equivalence class of functions having the same value and the same first derivatives at some designated point of the domain.

- Point transformations, of the form $x' = x'(x, z)$, $z' = z'(x, z)$; here, we think of (x, z, p) as coordinates on the space of tangent hyperplanes

$$\{dz - p_i dx^i\}^\perp \subset T_{(x^i, z)}(\mathbf{R}^{n+1})$$

of the manifold $\mathbf{R}^{n+1}$ with coordinates $(x^1, \ldots, x^n, z)$.

- Contact transformations, of the form $x' = x'(x, z, p)$, $z' = z'(x, z, p)$, $p' = p'(x, z, p)$, satisfying the equation of differential 1-forms

$$dz' - \sum p_i' dx^{i'} = f \cdot (dz - \sum p_i dx^i)$$

for some function $f(x, z, p) \neq 0$.

We will be studying the geometry of functionals $\mathcal{F}_L(z)$ subject to the class of contact transformations, which is strictly larger than the other three classes. The effects of this choice will become clear as we proceed. Although contact transformations were recognized classically, appearing most notably in studies of surface geometry, they do not seem to have been extensively utilized in the calculus of variations.

Classical calculus of variations primarily concerns the following features of a functional $\mathcal{F}_L$.

The *first variation* $\delta\mathcal{F}_L(z)$ is analogous to the derivative of a function, where $z = z(x)$ is thought of as an independent variable in an infinite-dimensional space of functions. The analog of the condition that a point be critical is the condition that $z(x)$ be *stationary* for all fixed-boundary variations. Formally, one writes

$$\delta\mathcal{F}_L(z) = 0,$$

and as we shall explain, this gives a second-order scalar partial differential equation for the unknown function $z(x)$ of the form

$$\frac{\partial L}{\partial z} - \sum \frac{d}{dx^i}\left(\frac{\partial L}{\partial p_i}\right) = 0.$$

This is the *Euler-Lagrange equation* of the Lagrangian $L(x, z, p)$, and we will study it in an invariant, geometric setting. This seems especially promising in light of the fact that, although it is not obvious, the process by which we associate an Euler-Lagrange equation to a Lagrangian is invariant under the large class of contact transformations. Also, note that the Lagrangian L determines the functional $\mathcal{F}_L$, but not vice versa. To see this, observe that if we add to $L(x, z, p)$ a "divergence term" and consider

$$L'(x, z, p) = L(x, z, p) + \sum\left(\frac{\partial K^i(x, z)}{\partial x^i} + \frac{\partial K^i(x, z)}{\partial z}p^i\right)$$

for functions $K^i(x, z)$, then by Green's theorem, the functionals $\mathcal{F}_L$ and $\mathcal{F}_{L'}$ differ by a constant depending only on values of z on $\partial\Omega$. For many purposes, such functionals should be considered equivalent; in particular, L and L' have the same Euler-Lagrange equations.

Second, there is a relationship between *symmetries* of a Lagrangian L and *conservation laws* for the corresponding Euler-Lagrange equations, described by a classical theorem of Noether. A subtlety here is that the group of symmetries of an equivalence class of Lagrangians may be strictly larger than the group of symmetries of any particular representative. We will investigate how this discrepancy is reflected in the space of conservation laws, in a manner that involves global topological issues.

Third, one considers the *second variation* $\delta^2 \mathcal{F}_L$, analogous to the Hessian of a smooth function, usually with the goal of identifying local minima of the functional. There has been a great deal of analytic work done in this area for classical variational problems, reducing the problem of local minimization to understanding the behavior of certain *Jacobi operators*, but the geometric theory is not as well-developed as that of the first variation and the Euler-Lagrange equations.

We will consider these issues and several others in a geometric setting as suggested above, using various methods from the subject of exterior differential systems, to be explained along the way. Chapter 1 begins with an introduction to *contact manifolds*, which provide the geometric setting for the study of first-order functionals (1) subject to contact transformations. We then construct an object that is central to the entire theory: the *Poincaré-Cartan form*, an explicitly computable differential form that is associated to the equivalence class of any Lagrangian, where the notion of equivalence includes that alluded to above for classical Lagrangians. We then carry out a calculation using the Poincaré-Cartan form to associate to any Lagrangian on a contact manifold an exterior differential system—the *Euler-Lagrange system*—whose integral manifolds are *stationary* for the associated functional; in the classical case, these correspond to solutions of the Euler-Lagrange equation. The Poincaré-Cartan form also makes it quite easy to state and prove *Noether's theorem*, which gives an isomorphism between a space of symmetries of a Lagrangian and a space of conservation laws for the Euler-Lagrange equation; the subject of exterior differential systems provides a particularly natural setting for studying the latter objects. We illustrate all of this theory in the case of minimal hypersurfaces in Euclidean space $\mathbf{E}^n$, and in the case of more general linear Weingarten surfaces in $\mathbf{E}^3$, providing intuitive and computationally simple proofs of known results.

In Chapter 2, we consider the geometry of Poincaré-Cartan forms more closely. The main tool for this is É. Cartan's *method of equivalence*, by which one develops an algorithm for associating to certain geometric structures their differential invariants under a specified class of equivalences. We explain the various steps of this method while illustrating them in several major cases. First, we apply the method to *hyperbolic Monge-Ampere systems* in two independent variables; these exterior differential systems include many important Euler-Lagrange systems that arise from classical problems, and among other results, we find a characterization of those PDEs that are contact-equivalent to the homogeneous linear wave equation. We then turn to the case of $n \geq 3$ independent variables, and carry out several steps of the equivalence method for Poincaré-Cartan forms, after isolating those of the algebraic type arising from classical problems. Associated to such a *neo-classical form* is a field of hypersurfaces in the fibers of a vector bundle, well-defined up to affine transformations. This motivates a digression on the affine geometry of hypersurfaces, conducted using Cartan's *method of moving frames*, which we will illustrate but not discuss in any generality. After identifying a number of differential invariants for Poincaré-Cartan forms in this manner, we show that they are sufficient for characterizing those Poincaré-Cartan forms associated to the PDE for hypersurfaces having prescribed mean curvature.

A particularly interesting branch of the equivalence problem for neo-classical Poincaré-Cartan forms includes some highly symmetric Poincaré-Cartan forms corresponding to Poisson equations, discussed in Chapter 3. Some of these equations have good invariance properties under the group of conformal transformations of the n-sphere, and we find that the corresponding branch of the equivalence problem reproduces a construction that is familiar in conformal geometry. We will discuss the relevant aspects of conformal geometry in some detail; these include another application of the equivalence method, in which the important conceptual step of *prolongation* of G-structures appears for the first time. This point of view allows us to apply Noether's theorem in a particularly simple way to the most symmetric of non-linear Poisson equations, the one with the critical exponent:

$$\Delta u = C u^{\frac{n+2}{n-2}}.$$

Having calculated the conservation laws for this equation, we also consider the case of wave equations, and in particular the very symmetric example:

$$\Box z = C z^{\frac{n+3}{n-1}}.$$

Here, conformal geometry with Lorentz signature is the appropriate background, and we present the conservation laws corresponding to the associated symmetry group, along with a few elementary applications.

The final chapter addresses certain matters which are thus far not so well-developed. First, we consider the second variation of a functional, with the goal of understanding which integral manifolds of an Euler-Lagrange system are local minima. We give an interesting geometric formula for the second variation, in which conformal geometry makes another appearance (unrelated to that in the preceding chapter). Specifically, we find that the critical submanifolds for certain variational problems inherit a canonical conformal structure, and the second variation can be expressed in terms of this structure and an additional scalar curvature invariant. This interpretation does not seem to appear in the classical literature. Circumstances under which one can carry out in an invariant manner the usual "integration by parts" in the second-variation formula, which is crucial for the study of local minimization, turn out to be somewhat limited.

We discuss the reason for this, and illustrate the optimal situation by revisiting the example of prescribed mean curvature systems.

We also consider the problem of finding an analog of the Poincaré-Cartan form in the case of functionals on vector-valued functions and their Euler-Lagrange PDE systems. Although there is no analog of proper contact transformations in this case, we will present and describe the merits of D. Betounes' construction of such an analog, based on some rather involved multi-linear algebra. An illuminating special case is that of harmonic maps between Riemannian manifolds, for which we find the associated forms and conservation laws.

Finally, we consider the appearance of higher-order conservation laws for first-order variational problems. The geometric setting for these is the *infinite prolongation* of an Euler-Lagrange system, which has come to play a major role in classifying conservation laws. We will propose a generalized version of Noether's theorem appropriate to our setting, but we do not have a proof of our statement. In any case, there are other ways to illustrate two of the most well-known but intriguing examples: the system describing Euclidean surfaces of Gauss curvature $K = -1$, and that corresponding to the sine-Gordon equation, $\Box z = \sin z$. We will generate examples of higher-order conservation laws by relating these two systems, first in the classical manner, and then more systematically using the notions of *prolongation* and *integrable extension*, which come from the subject of exterior differential systems. Finally, having explored these systems this far, it is convenient to exhibit and relate the *Bäcklund transformations* that act on each.

One particularly appealing aspect of this study is that one sees in action so many components of the subject of exterior differential systems. There are particularly beautiful instances of the method of equivalence, a good illustration of the method of moving frames (for affine hypersurfaces), essential use of prolongation both of G-structures and of differential systems, and a use of the notion of integrable extension to clarify a confusing issue.

Of course, the study of Euler-Lagrange equations by means of exterior differential forms and the method of equivalence is not new. In fact, much of the 19th century material in this area is so naturally formulated in terms of differential forms (cf. the Hilbert form in the one-variable calculus of variations) that it is difficult to say exactly when this approach was initiated.

However, there is no doubt that Élie Cartan's 1922 work *Leçons sur les invariants intégraux* [Car71] serves both as an elegant summary of the known material at the time and as a remarkably forward-looking formulation of the use of differential forms in the calculus of variations. At that time, Cartan did not bring his method of equivalence (which he had developed beginning around 1904 as a tool to study the geometry of pseudogroups) to bear on the subject. It was not until his 1933 work *Les espaces métriques fondés sur la notion d'aire* [Car33] and his 1934 monograph *Les espaces de Finsler* [Car34] that Cartan began to explore the geometries that one could attach to a Lagrangian for surfaces or for curves. Even in these works, any explicit discussion of the full method of equivalence is supressed and Cartan contents himself with deriving the needed geometric structures by seemingly ad hoc methods.

After the modern formulation of jet spaces and their contact systems was put into place, Cartan's approach was extended and further developed by several people. One might particularly note the 1935 work of Th. de Donder [Don35] and its development. Beginning in the early 1940s, Th. Lepage [Lep46, Lep54] undertook a study of first-order Lagrangians that made extensive use of the algebra of differential forms on a contact manifold. Beginning in the early 1950s, this point of view was developed further by P. Dedecker [Ded77], who undertook a serious study of the calculus of variations via tools of homological algebra. All of these authors are concerned in one way or another with the canonical construction of differential geometric (and other) structures associated to a Lagrangian, but the method of equivalence is not utilized in any extensive way. Consequently, they deal primarily with first-order linear-algebraic invariants of variational problems. Only with the method of equivalence can one uncover the full set of higher-order geometric invariants. This is one of the central themes of the present work; without the equivalence method, for example, one could not give our unique characterizations of certain classical, "natural" systems (cf. §2.1, §2.5, and §3.3).

In more modern times, numerous works of I. Anderson, D. Betounes, R. Hermann, N. Kamran, V. Lychagin, P. Olver, H. Rund, A. Vinogradov, and their coworkers, just to name a few, all concern themselves with geometric aspects and invariance properties of the calculus of variations. Many of the results expounded in this monograph can be found in one form or another in works by these or earlier authors. We certainly make no pretense of giving a complete historical account of the work in this area in the 20th century. Our bibliography lists those works of which we were aware that seemed most relevant to our approach, if not necessarily to the results themselves, and it identifies only a small portion of the work done in these areas. The most substantially developed alternative theory in this area is that of the *variational bicomplex* associated to the algebra of differential forms on a fiber bundle. The reader can learn this material from Anderson's works [And92] and [And], and references therein, which contain results heavily overlapping those of our Chapter 4.

Some terminology and notation that we will use follows, with more introduced in the text. An *exterior differential system* (EDS) is a pair $(M, \mathcal{E})$ consisting of a smooth manifold M and a homogeneous, differentially closed ideal $\mathcal{E} \subseteq \Omega^*(M)$ in the algebra of smooth differential forms on M. Some of the EDSs that we study are differentially generated by the sections of a smooth subbundle $I \subseteq T^*M$ of the cotangent bundle of M; this subbundle, and sometimes its space of sections, is called a *Pfaffian system* on M. It will be useful to have the notation $\{\alpha, \beta, \ldots\}$ for the (two-sided) *algebraic* ideal generated by forms $\alpha, \beta, \ldots$, and to have the notation $\{I\}$ for the algebraic ideal generated by the sections of a Pfaffian system $I \subseteq T^*M$. An *integral manifold* of an EDS $(M, \mathcal{E})$ is a submanifold immersion $\iota : N \hookrightarrow M$ for which $\varphi_N \stackrel{\text{def}}{=} \iota^*\varphi = 0$ for all $\varphi \in \mathcal{E}$. Integral manifolds of Pfaffian systems are defined similarly.

A differential form φ on the total space of a fiber bundle $\pi : E \to B$ is said to be *semibasic* if its contraction with any vector field tangent to the fibers of π vanishes, or equivalently, if its value at each point $e \in E$ is the pullback via π_e^* of some form at $\pi(e) \in B$. Some authors call such a form *horizontal*. A stronger condition is that φ be *basic*, meaning that it is locally (in open subsets of E) the pullback via π^* of a form on the base B.

Our computations will frequently require the following multi-index notation. If $(\omega^1, \ldots, \omega^n)$ is an ordered basis for a vector space V, then corresponding to a multi-index $I = (i_1, \ldots, i_k)$ is the k-vector

$$\omega^I = \omega^{i_1} \wedge \cdots \wedge \omega^{i_k} \in \textstyle\bigwedge^k(V),$$

and for the complete multi-index we simply define

$$\omega = \omega^1 \wedge \cdots \wedge \omega^n.$$

Letting $(e_1, \ldots, e_n)$ be a dual basis for V^*, we also define the $(n-k)$-vector

$$\omega_{(I)} = e_I \lrcorner \, \omega = e_{i_k} \lrcorner \, (e_{i_{k-1}} \lrcorner \cdots (e_{i_1} \lrcorner \, \omega) \cdots).$$

This $\omega_{(I)}$ is, up to sign, just ω^{I_c}, where I_c is a multi-index complementary to I. For the most frequently occurring cases $k = 1, 2$ we have the formulae (with "hats" ˆ indicating omission of a factor)

$$
\begin{aligned}
\omega_{(i)} &= (-1)^{i-1} \omega^1 \wedge \cdots \wedge \hat{\omega}^i \wedge \cdots \wedge \omega^n, \\
\omega_{(ij)} &= (-1)^{i+j-1} \omega^1 \wedge \cdots \wedge \hat{\omega}^i \wedge \cdots \wedge \hat{\omega}^j \wedge \cdots \wedge \omega^n \\
&= -\omega_{(ji)}, \quad \text{for } i < j,
\end{aligned}
$$

and the identities

$$
\begin{aligned}
\omega^i \wedge \omega_{(j)} &= \delta_j^i \omega, \\
\omega^i \wedge \omega_{(jk)} &= \delta_k^i \omega_{(j)} - \delta_j^i \omega_{(k)}.
\end{aligned}
$$

We will often, but not always, use without comment the convention of summing over repeated indices. Always, $n \geq 2$.[2]

---

[2] For the case $n = 1$, an analogous geometric approach to the calculus of variations for curves may be found in [Gri83].

Chapter 1

Lagrangians and Poincaré-Cartan Forms

In this chapter, we will construct and illustrate our basic objects of study. The geometric setting that one uses for studying Lagrangian functionals subject to contact transformations is a *contact manifold*, and we will begin with its definition and relevant cohomological properties. These properties allow us to formalize an intuitive notion of equivalence for functionals, and more importantly, to replace such an equivalence class by a more concrete differential form, the *Poincaré-Cartan form*, on which all of our later calculations depend. In particular, we will first use it to derive the Euler-Lagrange differential system, whose integral manifolds correspond to stationary points of a given functional. We then use it to give an elegant version of the solution to the inverse problem, which asks when a differential system of the appropriate algebraic type is the Euler-Lagrange system of some functional. Next, we use it to define the isomorphism between a certain Lie algebra of infinitesimal symmetries of a variational problem and the space conservation laws for the Euler-Lagrange system, as described in Noether's theorem. All of this will be illustrated at an elementary level using examples from Euclidean hypersurface geometry.

1.1 Lagrangians and Contact Geometry

We begin by introducing the geometric setting in which we will study Lagrangian functionals and their Euler-Lagrange systems.

Definition 1.1 *A* contact manifold (M, I) *is a smooth manifold M of dimension $2n + 1$ ($n \in \mathbf{Z}^+$), with a distinguished line subbundle $I \subset T^*M$ of the cotangent bundle which is non-degenerate in the sense that for any local 1-form θ generating I,*

$$\theta \wedge (d\theta)^n \neq 0.$$

9

Note that the non-degeneracy criterion is independent of the choice of θ; this is because if $\bar{\theta} = f\theta$ for some function $f \neq 0$, then we find

$$\bar{\theta} \wedge (d\bar{\theta})^n = f^{n+1} \theta \wedge (d\theta)^n.$$

For example, on the space $J^1(\mathbf{R}^n, \mathbf{R})$ of 1-jets of functions, we can take coordinates (x^i, z, p_i) corresponding to the jet at $(x^i) \in \mathbf{R}^n$ of the linear function $f(\bar{x}) = z + \sum p_i(\bar{x}^i - x^i)$. Then we define the *contact form*

$$\theta = dz - \sum p_i dx^i,$$

for which

$$d\theta = -\sum dp_i \wedge dx^i,$$

so the non-degeneracy condition $\theta \wedge (d\theta)^n \neq 0$ is apparent. In fact, the Pfaff theorem (cf. Ch. I, §3 of [B+91]) implies that every contact manifold is locally isomorphic to this example; that is, every contact manifold (M, I) has local coordinates (x^i, z, p_i) for which the form $\theta = dz - \sum p_i dx^i$ generates I.

More relevant for differential geometry is the example $G_n(TX^{n+1})$, the Grassmannian bundle parameterizing n-dimensional oriented subspaces of the tangent spaces of an $(n+1)$-dimensional manifold X. It is naturally a contact manifold, and will be considered in more detail later.

Let (M, I) be a contact manifold of dimension $2n + 1$, and assume that I is generated by a global, non-vanishing section $\theta \in \Gamma(I)$; this assumption only simplifies our notation, and would in any case hold on a double cover of M. Sections of I generate the *contact differential ideal*

$$\mathcal{I} = \{\theta, d\theta\} \subset \Omega^*(M)$$

in the exterior algebra of differential forms on M.[1] A *Legendre submanifold* of M is an immersion $\iota : N \hookrightarrow M$ of an n-dimensional submanifold N such that $\iota^*\theta = 0$ for any contact form $\theta \in \Gamma(I)$; in this case $\iota^* d\theta = 0$ as well, so a Legendre submanifold is the same thing as an integral manifold of the differential ideal $\mathcal{I}$. In Pfaff coordinates with $\theta = dz - \sum p_i dx^i$, one such integral manifold is given by

$$N_0 = \{z = p_i = 0\}.$$

To see other Legendre submanifolds "near" this one, note than any submanifold C^1-close to N_0 satisfies the independence condition

$$dx^1 \wedge \cdots \wedge dx^n \neq 0,$$

and can therefore be described locally as a graph

$$N = \{(x^i, z(x), p_i(x))\}.$$

---

[1] Recall our convention that braces $\{\cdot\}$ denote the algebraic ideal generated by an object; for instance, $\{\theta\}$ consists of exterior multiples of any contact form θ, and is smaller than $\mathcal{I}$. We sometimes use $\{I\}$ as alternate notation for $\{\theta\}$.

In this case, we have

$$\theta|_N = 0 \quad \text{if and only if} \quad p_i(x) = \frac{\partial z}{\partial x^i}(x).$$

Therefore, N is determined by the function $z(x)$, and conversely, every function $z(x)$ determines such an N; we informally say that "the generic Legendre submanifold depends locally on one arbitrary function of n variables." Legendre submanifolds of this form, with $dx|_N \neq 0$, will often be described as *transverse*.

Motivated by (1) in the Introduction, we are primarily interested in functionals given by triples (M, I, Λ), where (M, I) is a $(2n + 1)$-dimensional contact manifold, and $\Lambda \in \Omega^n(M)$ is a differential form of degree n on M; such a Λ will be referred to as a *Lagrangian* on (M, I).[2] We then define a functional on the set of smooth, compact Legendre submanifolds $N \subset M$, possibly with boundary ∂N, by

$$\mathcal{F}_\Lambda(N) = \int_N \Lambda.$$

The classical variational problems described above may be recovered from this notion by taking $M = J^1(\mathbf{R}^n, \mathbf{R}) \cong \mathbf{R}^{2n+1}$ with coordinates (x^i, z, p_i), I generated by $\theta = dz - \sum p_i dx^i$, and $\Lambda = L(x^i, z, p_i) dx$. This formulation also admits certain functionals depending on second derivatives of $z(x)$, because there may be dp_i terms in Λ. Later, we will restrict attention to a class of functionals which, possibly after a contact transformation, can be expressed without second derivatives.

There are two standard notions of equivalence for Lagrangians Λ. First, note that if the difference $\Lambda - \Lambda'$ of two Lagrangians lies in the contact ideal $\mathcal{I}$ then the functionals $\mathcal{F}_\Lambda$ and $\mathcal{F}_{\Lambda'}$ are equal, because they are defined only for Legendre submanifolds, on which all forms in $\mathcal{I}$ vanish. Second, suppose that the difference of two Lagrangians is an exact n-form, $\Lambda - \Lambda' = d\varphi$ for some $\varphi \in \Omega^{n-1}(M)$. Then we find

$$\mathcal{F}_\Lambda(N) = \mathcal{F}_{\Lambda'}(N) + \int_{\partial N} \varphi$$

for all Legendre submanifolds N. One typically studies the variation of $\mathcal{F}_\Lambda$ along 1-parameter families N_t with fixed boundary, and the preceding equation shows that $\mathcal{F}_\Lambda$ and $\mathcal{F}_{\Lambda'}$ differ only by a constant on such a family. Such Λ and Λ' are sometimes said to be *divergence-equivalent*.

These two notions of equivalence suggest that we consider the class

$$[\Lambda] \in \Omega^n(M)/(\mathcal{I}^n + d\Omega^{n-1}(M)),$$

where $\mathcal{I}^n = \mathcal{I} \cap \Omega^n(M)$. The natural setting for this space is the quotient $(\bar{\Omega}^*, \bar{d})$ of the de Rham complex $(\Omega^*(M), d)$, where $\bar{\Omega}^n = \Omega^n(M)/\mathcal{I}^n$, and $\bar{d}$ is induced by the usual exterior derivative d on this quotient. We then have

---

[2] In the Introduction, we used the term *Lagrangian* for a function, rather than for a differential form, but we will not do so again.

characteristic cohomology groups $\bar{H}^n = H^n(\bar{\Omega}^*, \bar{d})$. We will show in a moment that (recalling $\dim(M) = 2n + 1$):

$$\text{for } k > n, \quad \mathcal{I}^k = \Omega^k(M). \tag{1.1}$$

In other words, all forms on M of degree greater than n lie in the contact ideal; one consequence is that $\mathcal{I}$ can have no integral manifolds of dimension greater than n. The importance of (1.1) is that it implies that $d\Lambda \in \mathcal{I}^{n+1}$, and we can therefore regard our equivalence class of functionals as a characteristic cohomology class

$$[\Lambda] \in \bar{H}^n.$$

This class is almost, but not quite, our fundamental object of study.

To prove both (1.1) and several later results, we need to describe some of the pointwise linear algebra associated with the contact ideal $\mathcal{I} = \{\theta, d\theta\} \subset \Omega^*(M)$. Consider the tangent distribution of rank $2n$

$$I^\perp \subset TM$$

given by the annihilator of the contact line bundle. Then the non-degeneracy condition on I implies that the 2-form

$$\Theta \overset{def}{=} d\theta$$

restricts fiberwise to $I^\perp$ as a non-degenerate, alternating bilinear form, determined by I up to scaling. This allows one to use tools from symplectic linear algebra; the main fact is the following.

Proposition 1.1 *Let* (V^{2n}, Θ) *be a symplectic vector space, where* $\Theta \in \bigwedge^2 V^*$ *is a non-degenerate alternating bilinear form. Then*
(a) for $0 \le k \le n$, *the map*

$$(\Theta\wedge)^k : \bigwedge^{n-k} V^* \to \bigwedge^{n+k} V^* \tag{1.2}$$

is an isomorphism, and
(b) if we define the space of primitive forms *to be*

$$P^{n-k}(V^*) = \text{Ker}\ ((\Theta\wedge)^{k+1} : \bigwedge^{n-k} V^* \to \bigwedge^{n+k+2} V^*),$$

then we have a decomposition of $Sp(n, \mathbf{R})$*-modules*

$$\bigwedge^{n-k} V^* = P^{n-k}(V^*) \oplus \left(\Theta \wedge \bigwedge^{n-k-2} V^*\right).\text{[3]}$$

---

[3] *Of course, we can extend this decomposition inductively to obtain the* Hodge-Lepage *decomposition*

$$\bigwedge^{n-k} V^* \cong P^{n-k}(V^*) \oplus P^{n-k-2}(V^*) \oplus \cdots \oplus \begin{cases} P^1(V^*) = V^* \text{ for } n-k \text{ odd,} \\ P^0(V^*) = \mathbf{R} \text{ for } n-k \text{ even,} \end{cases}$$

Proposition 1.1 implies in particular (1.1), for it says that modulo $\{\theta\}$ (equivalently, restricted to $I^\perp$), every form φ of degree greater than n is a multiple of $d\theta$, which is exactly to say that φ is in the algebraic ideal generated by θ and $d\theta$.

Proof. (a) Because $\bigwedge^{n-k} V^*$ and $\bigwedge^{n+k} V^*$ have the same dimension, it suffices to show that the map (1.2) is injective. We proceed by induction on k, downward from $k = n$ to $k = 0$. In case $k = n$, the (1.2) is just multiplication

$$(\Theta^n)\cdot : \mathbf{R} \to \textstyle\bigwedge^{2n} V^*,$$

which is obviously injective, because Θ is non-degenerate.

Now suppose that the statement is proved for some k, and suppose that $\xi \in \bigwedge^{n-(k-1)}$ satisfies

$$\Theta^{k-1} \wedge \xi = 0.$$

This implies that

$$\Theta^k \wedge \xi = 0,$$

so that for every vector $X \in V$, we have

$$0 = X \lrcorner (\Theta^k \wedge \xi) = k(X \lrcorner \Theta) \wedge \Theta^{k-1} \wedge \xi + \Theta^k \wedge (X \lrcorner \xi).$$

Now, the first term on the right-hand side vanishes by our assumption on ξ (our second use of this assumption), so we must have

$$0 = \Theta^k \wedge (X \lrcorner \xi),$$

and the induction hypothesis then gives

$$X \lrcorner \xi = 0.$$

This is true for every $X \in V$, so we conclude that $\xi = 0$.

(b) We will show that any $\xi \in \bigwedge^{n-k} V^*$ has a unique decomposition as the sum of a primitive form and a multiple of Θ. For the existence of such a decomposition, we apply the surjectivity in part (a) to the element $\Theta^{k+1} \wedge \xi \in \bigwedge^{n+k+2} V^*$, and find $\eta \in \bigwedge^{n-k-2} V^*$ for which

$$\Theta^{k+2} \wedge \eta = \Theta^{k+1} \wedge \xi.$$

under which any element $\xi \in \bigwedge^{n-k} V^*$ can be written uniquely as

$$\xi = \xi_0 + (\Theta \wedge \xi_1) + (\Theta^2 \wedge \xi_2) + \cdots + \left(\Theta^{\lfloor \frac{n-k}{2} \rfloor} \wedge \xi_{\lfloor \frac{n-k}{2} \rfloor}\right),$$

with each $\xi_i \in P^{n-k-2i}(V^*)$. What we will not prove here is that the representation of $Sp(n, \mathbf{R})$ on $P^{n-k}(V^*)$ is irreducible for each k, so this gives the complete irreducible decomposition of $\bigwedge^{n-k}(V^*)$.

Then we can decompose

$$\xi = (\xi - \Theta \wedge \eta) + (\Theta \wedge \eta),$$

where the first summand is primitive by construction.

To prove uniqueness, we need to show that if $\Theta \wedge \eta$ is primitive for some $\eta \in \bigwedge^{n-k-2}(V^*)$, then $\Theta \wedge \eta = 0$. In fact, primitivity means

$$0 = \Theta^k \wedge \Theta \wedge \eta,$$

which implies that $\eta = 0$ by the injectivity in part (a).   □

Returning to our discussion of Lagrangian functionals, observe that there is a short exact sequence of complexes

$$0 \to \mathcal{I}^* \to \Omega^*(M) \to \bar{\Omega}^* \to 0$$

giving a long exact cohomology sequence

$$\cdots \to H_{dR}^n(M) \to \bar{H}^n \overset{\delta}{\to} H^{n+1}(\mathcal{I}) \to H_{dR}^{n+1}(M) \to \cdots,$$

where δ is essentially exterior differentiation. Although an equivalence class $[\Lambda] \in \bar{H}^n$ generally has no canonical representative differential form, we can now show that its image $\delta([\Lambda]) \in H^{n+1}(\mathcal{I})$ does.

Theorem 1.1 *Any class* $[\Pi] \in H^{n+1}(\mathcal{I})$ *has a unique global representative closed form* $\Pi \in \mathcal{I}^{n+1}$ *satisfying* $\theta \wedge \Pi = 0$ *for any contact form* $\theta \in \Gamma(I)$, *or equivalently,* $\Pi \equiv 0$ *(mod* $\{I\}$).

Proof. Any $\Pi \in \mathcal{I}^{n+1}$ may be written locally as

$$\Pi = \theta \wedge \alpha + d\theta \wedge \beta$$

for some $\alpha \in \Omega^n(M)$, $\beta \in \Omega^{n-1}(M)$. But this is the same as

$$\Pi = \theta \wedge (\alpha + d\beta) + d(\theta \wedge \beta),$$

so replacing Π with the equivalent (in $H^{n+1}(\mathcal{I})$) form $\Pi - d(\theta \wedge \beta)$, we have the *local* existence of a representative as claimed.

For uniqueness, suppose that $\Pi_1 - \Pi_2 = d(\theta \wedge \gamma)$ for some $(n-1)$-form γ (this is exactly equivalence in $H^{n+1}(\mathcal{I})$), and that $\theta \wedge \Pi_1 = \theta \wedge \Pi_2 = 0$. Then $\theta \wedge d\theta \wedge \gamma = 0$, so $d\theta \wedge \gamma \equiv 0$ (mod $\{I\}$). By symplectic linear algebra, this implies that $\gamma \equiv 0$ (mod $\{I\}$), so $\Pi_1 - \Pi_2 = 0$.

Finally, global existence follows from local existence and uniqueness.   □
We can now define our main object of study.

Definition 1.2 *For a contact manifold* (M, I) *with Lagrangian* Λ, *the unique representative* $\Pi \in \mathcal{I}^{n+1}$ *of* $\delta([\Lambda])$ *satisfying* $\Pi \equiv 0$ *(mod* $\{I\}$) *is called the* Poincaré-Cartan form *of* Λ.

Poincaré-Cartan forms of Lagrangians will be the main object of study in these lectures, and there are two computationally useful ways to think of them. The first is as above: given a representative Lagrangian Λ, express $d\Lambda$ locally as $\theta \wedge (\alpha + d\beta) + d(\theta \wedge \beta)$, and then

$$\boxed{\Pi = \theta \wedge (\alpha + d\beta).}$$

The second, which will be important for computing the first variation and the Euler-Lagrange system of $[\Lambda]$, is as an exact form:

$$\boxed{\Pi = d(\Lambda - \theta \wedge \beta).}$$

In fact, β is the unique $(n-1)$-form modulo $\{I\}$ such that

$$d(\Lambda - \theta \wedge \beta) \equiv 0 \pmod{\{I\}}.$$

This observation will be used later, in the proof of Noether's theorem.

1.2 The Euler-Lagrange System

In the preceding section, we showed how one can associate to an equivalence class $[\Lambda]$ of Lagrangians on a contact manifold (M, I) a canonical $(n+1)$-form Π. In this section, we use this Poincaré-Cartan form to find an exterior differential system whose integral manifolds are precisely the stationary Legendre submanifolds for the functional $\mathcal{F}_\Lambda$. This requires us to calculate the *first variation* of $\mathcal{F}_\Lambda$, which gives the derivative of $\mathcal{F}_\Lambda(N_t)$ for any 1-parameter family N_t of Legendre submanifolds of (M, I). The Poincaré-Cartan form enables us to carry out the usual integration by parts for this calculation in an invariant manner.

We also consider the relevant version of the *inverse problem* of the calculus of variations, which asks whether a given PDE of the appropriate type is equivalent to the Euler-Lagrange equation for some functional. We answer this by giving a necessary and sufficient condition for an EDS of the appropriate type to be locally equivalent to the Euler-Lagrange system of some $[\Lambda]$. We find these conditions by reducing the problem to a search for a Poincaré-Cartan form.

1.2.1 Variation of a Legendre Submanifold

Suppose that we have a 1-parameter family $\{N_t\}$ of Legendre submanifolds of a contact manifold (M, I); more precisely, this is given by a compact manifold with boundary $(N, \partial N)$ and a smooth map

$$F : N \times [0, 1] \to M$$

which is a Legendre submanifold F_t for each fixed $t \in [0, 1]$ and is independent of $t \in [0, 1]$ on $\partial N \times [0, 1]$. Because $F_t^* \theta = 0$ for any contact form $\theta \in \Gamma(I)$, we must have locally

$$F^* \theta = G \, dt \tag{1.3}$$

for some function G on $N \times [0,1]$. We let $g = G|_{N \times \{0\}}$ be the restriction to the initial submanifold.

It will be useful to know that given a Legendre submanifold $f : N \hookrightarrow M$, every function g may be realized as in (1.3) for some fixed-boundary variation and some contact form θ, locally in the interior N°. This may be seen in Pfaff coordinates (x^i, z, p_i) on M, for which $\theta = dz - \sum p_i dx^i$ generates I and such that our given N is a 1-jet graph $\{(x^i, z(x), p_i(x) = z_{x^i}(x))\}$. Then (x^i) give coordinates on N, and a variation of N is of the form

$$F(x,t) = (x^i, z(x,t), z_{x^i}(x,t)).$$

Now $F^*(dz - \sum p_i dx^i) = z_t dt$; and given $z(x,0)$, we can always extend to $z(x,t)$ with $g(x) = z_t(x,0)$ prescribed arbitrarily, which is what we claimed.

1.2.2   Calculation of the Euler-Lagrange System

We can now carry out a calculation that is fundamental for the whole theory. Suppose given a Lagrangian $\Lambda \in \Omega^n(M)$ on a contact manifold (M,I), and a fixed-boundary variation of Legendre submanifold $F : N \times [0,1] \to M$; we wish to compute $\frac{d}{dt}(\int_{N_t} \Lambda)$.

To do this, first recall the calculation of the Poincaré-Cartan form for the equivalence class $[\Lambda] \in \bar{H}^n$. Because $\mathcal{I}^{n+1} = \Omega^{n+1}(M)$, we can always write

$$
\begin{aligned}
d\Lambda &= \theta \wedge \alpha + d\theta \wedge \beta \\
&= \theta \wedge (\alpha + d\beta) + d(\theta \wedge \beta),
\end{aligned}
$$

and then

$$\Pi = \theta \wedge (\alpha + d\beta) = d(\Lambda - \theta \wedge \beta). \tag{1.4}$$

We are looking for conditions on a Legendre submanifold $f : N \hookrightarrow M$ to be *stationary* for $[\Lambda]$ under all fixed-boundary variations, in the sense that $\frac{d}{dt}\big|_{t=0}(\int_{N_t} \Lambda) = 0$ whenever $F|_{t=0} = f$. We compute (without writing the F^*s)

$$
\begin{aligned}
\frac{d}{dt}\int_{N_t} \Lambda &= \frac{d}{dt}\int_{N_t}(\Lambda - \theta \wedge \beta) \\
&= \int_{N_t} \mathcal{L}_{\frac{\partial}{\partial t}}(\Lambda - \theta \wedge \beta) \\
&= \int_{N_t}\left(\tfrac{\partial}{\partial t} \lrcorner\, d(\Lambda - \theta \wedge \beta)\right) + \int_{N_t} d\left(\tfrac{\partial}{\partial t} \lrcorner (\Lambda - \theta \wedge \beta)\right) \\
&= \int_{N_t} \tfrac{\partial}{\partial t} \lrcorner\, \Pi \quad \text{(because } \partial N \text{ is fixed).}
\end{aligned}
$$

One might express this result as

$$\delta(\mathcal{F}_\Lambda)_N(v) = \int_N v \lrcorner\, f^*\Pi,$$

where the variational vector field v, lying in the space $\Gamma_0(f^*TM)$ of sections of f^*TM vanishing along ∂N, plays the role of $\frac{\partial}{\partial t}$. The condition $\Pi \equiv 0 \pmod{\{I\}}$ allows us to write $\Pi = \theta \wedge \Psi$ for some n-form Ψ, not uniquely determined, and we have

$$\frac{d}{dt}\bigg|_{t=0} \int_{N_t} \Lambda = \int_N g\, f^*\Psi,$$

where $g = (\frac{\partial}{\partial t} \, \lrcorner\, F^*\theta)|_{t=0}$. It was shown previously that this g could locally be chosen arbitrarily in the interior N^o, so the necessary and sufficient condition for a Legendre submanifold $f : N \hookrightarrow M$ to be stationary for $\mathcal{F}_\Lambda$ is that $f^*\Psi = 0$.

Definition 1.3 *The* Euler-Lagrange system *of the Lagrangian Λ is the differential ideal generated algebraically as*

$$\mathcal{E}_\Lambda = \{\theta, d\theta, \Psi\} \subset \Omega^*(M).$$

A stationary Legendre submanifold of Λ is an integral manifold of $\mathcal{E}_\Lambda$. The functional is said to be non-degenerate *if its Poincaré-Cartan form $\Pi = \theta \wedge \Psi$ has no degree-1 divisors (in the exterior algebra of T^*M) other than multiples of θ.*

Note first that $\mathcal{E}_\Lambda$ is uniquely determined by Π, even though θ and Ψ may not be.[4] Note also that the ideal in $\Omega^*(M)$ algebraically generated by $\{\theta, d\theta, \Psi\}$ is already differentially closed, because $d\Psi \in \Omega^{n+1}(M) = \mathcal{I}^{n+1}$.

We can examine this for the classical situation where $M = \{(x^i, z, p_i)\}$, $\theta = dz - \sum p_i dx^i$, and $\Lambda = L(x, z, p)dx$. We find

$$\begin{aligned}
d\Lambda &= L_z \theta \wedge dx + \sum L_{p_i} dp_i \wedge dx \\
&= \theta \wedge L_z dx - d\theta \wedge \sum L_{p_i} dx_{(i)},
\end{aligned}$$

so referring to (1.4),

$$\Pi = \theta \wedge \left(L_z dx - \sum d(L_{p_i} dx_{(i)}) \right) = \theta \wedge \Psi.$$

Now, for a transverse Legendre submanifold $N = \{(x^i, z(x), z_{x^i}(x))\}$, we have $\Psi|_N = 0$ if and only if along N

$$\frac{\partial L}{\partial z} - \sum \frac{d}{dx^i}\left(\frac{\partial L}{\partial p_i}\right) = 0,$$

---

[4] Actually, given Π we have not only a well-defined $\mathcal{E}_\Lambda$, but a well-defined Ψ modulo $\{I\}$ which is primitive on $I^\perp$. There is a canonical map $\mathcal{E} : H^n(\bar{\Omega}^*) \to P^n(T^*M/I)$ to the space of primitive forms, taking a Lagrangian class $[\Lambda]$ to the corresponding Ψ in its Euler-Lagrange system; and this map fits into a full resolution of the constant sheaf

$$0 \to \mathbf{R} \to \bar{\Omega}^0 \to \cdots \to \bar{\Omega}^{n-1} \to H^n(\bar{\Omega}^*) \xrightarrow{\mathcal{E}} P^n(T^*M/I) \to \cdots \to P^0(T^*M/I) \to 0.$$

This has been developed and applied in the context of CR geometry in [Rum90].

where

$$\frac{d}{dx^i} = \frac{\partial}{\partial x^i} + z_{x^i}\frac{\partial}{\partial z} + \sum_j z_{x^i x^j}\frac{\partial}{\partial p_j}$$

is the *total derivative*. This is the usual Euler-Lagrange equation, a second-order, quasi-linear PDE for $z(x^1, \ldots, x^n)$ having symbol $L_{p_i p_j}$. It is an exercise to show that this symbol matrix is invertible at (x^i, z, p_i) if and only if Λ is non-degenerate in the sense of Definition 1.3.

1.2.3 The Inverse Problem

There is a reasonable model for exterior differential systems of "Euler-Lagrange type."

Definition 1.4 *A* Monge-Ampere differential system $(M, \mathcal{E})$ *consists of a contact manifold* (M, I) *of dimension* $2n + 1$, *together with a differential ideal* $\mathcal{E} \subset \Omega^*(M)$, *generated locally by the contact ideal* $\mathcal{I}$ *and an* n-*form* $\Psi \in \Omega^n(M)$.

Note that in this definition, the contact line bundle I can be recovered from $\mathcal{E}$ as its degree-1 part. We can now pose a famous question.

Inverse Problem: *When is a given Monge-Ampere system* $\mathcal{E}$ *on* M *equal to the Euler-Lagrange system* $\mathcal{E}_\Lambda$ *of some Lagrangian* $\Lambda \in \Omega^n(M)$?

Note that if a given $\mathcal{E}$ does equal $\mathcal{E}_\Lambda$ for some Λ, then for some local generators θ, Ψ of $\mathcal{E}$ we must have $\theta \wedge \Psi = \Pi$, the Poincaré-Cartan form of Λ. Indeed, we can say that $(M, \mathcal{E})$ is Euler-Lagrange if and only if there is an exact form $\Pi \in \Omega^{n+1}(M)$, locally of the form $\theta \wedge \Psi$ for some generators θ, Ψ of $\mathcal{E}$. However, we face the difficulty that $(M, \mathcal{E})$ does not determine either $\Psi \in \Omega^n(M)$ or $\theta \in \Gamma(I)$ uniquely.

This can be partially overcome by normalizing Ψ as follows. Given only $(M, \mathcal{E} = \{\theta, d\theta, \Psi\})$, Ψ is determined as an element of $\bar{\Omega}^n = \Omega^n(M)/\mathcal{I}^n$. We can obtain a representative Ψ that is unique modulo $\{I\}$ by adding the unique multiple of $d\theta$ that yields a *primitive* form on $I^\perp$, referring to the symplectic decomposition of $\bigwedge^n(T^*M/I)$ (see Proposition 1.1). With this choice, we have a form $\theta \wedge \Psi$ which is uniquely determined up to scaling; the various multiples $f\theta \wedge \Psi$, where f is a locally defined function on M, are the candidates to be Poincaré-Cartan form. Note that using a primitive normalization is reasonable, because our actual Poincaré-Cartan forms $\Pi = \theta \wedge \Psi$ satisfy $d\Pi = 0$, which in particular implies that Ψ is primitive on $I^\perp$. The proof of Noether's theorem in the next section will use a more refined normalization of Ψ.

The condition for a Monge-Ampere system to be Euler-Lagrange is therefore that there should be a globally defined exact n-form Π, locally of the form $f\theta \wedge \Psi$ with Ψ normalized as above. This suggests the more accessible *local inverse problem*, which asks whether there is a *closed* n-form that is locally expressible as $f\theta \wedge \Psi$. It is for this local version that we give a criterion.

We start with any candidate Poincaré-Cartan form $\Xi = \theta \wedge \Psi$, and consider the following criterion on Ξ:

$$\boxed{d\Xi = \varphi \wedge \Xi \quad \text{for some } \varphi \text{ with } d\varphi \equiv 0 \ (\mathrm{mod} \ \mathcal{I}).} \tag{1.5}$$

We first note that if this holds for some choice of $\Xi = \theta \wedge \Psi$, then it holds for all other choices $f\Xi$; this is easily verified.

Second, we claim that if (1.5) holds, then we can find $\tilde{\varphi}$ also satisfying $d\Xi = \tilde{\varphi} \wedge \Xi$, and in addition, $d\tilde{\varphi} = 0$. To see this, write

$$d\varphi = \theta \wedge \alpha + \beta \, d\theta$$

(here α is a 1-form and β is a function), and differentiate using $d^2 = 0$, modulo the algebraic ideal $\{I\}$, to obtain

$$0 \equiv d\theta \wedge (\alpha + d\beta) \quad (\mathrm{mod} \ \{I\}).$$

But with the standing assumption $n \geq 2$, symplectic linear algebra implies that the 1-form $\alpha + d\beta$ must vanish modulo $\{I\}$. As a result,

$$d(\varphi - \beta \, \theta) = \theta \wedge (\alpha + d\beta) = 0,$$

so we can take $\tilde{\varphi} = \varphi - \beta \, \theta$, verifying the claim.

Third, once we know that $d\Xi = \varphi \wedge \Xi$ with $d\varphi = 0$, then on a possibly smaller neighborhood, we use the Poincaré lemma to write $\varphi = du$ for a function u, and then

$$d(e^{-u}\Xi) = e^{-u}(\varphi \wedge \Xi - du \wedge \Xi) = 0.$$

This proves the following.

Theorem 1.2 *A Monge-Ampere system* $(M, \mathcal{E} = \{\theta, d\theta, \Psi\})$ *on a* $(2n + 1)$-*dimensional contact manifold* M *with* $n \geq 2$, *where* Ψ *is assumed to be primitive modulo* $\{I\}$, *is locally equal to an Euler-Lagrange system* $\mathcal{E}_\Lambda$ *if and only if it satisfies (1.5).*

Example 1. Consider a scalar PDE of the form

$$\Delta z = f(x, z, \nabla z), \tag{1.6}$$

where $\Delta = \sum \frac{\partial^2}{\partial x^{i2}}$; we ask which functions $f : \mathbf{R}^{2n+1} \to \mathbf{R}$ are such that (1.6) is contact-equivalent to an Euler-Lagrange equation. To apply our framework, we let $M = J^1(\mathbf{R}^n, \mathbf{R})$, $\theta = dz - \sum p_i dx^i$ so $d\theta = -\sum dp_i \wedge dx^i$, and set

$$\Psi = \sum dp_i \wedge dx_{(i)} - f(x, z, p)dx.$$

Restricted to a Legendre submanifold of the form $N = \{(x^i, z(x), \frac{\partial z}{\partial x^i}(x)\}$, we find

$$\Psi|_N = (\Delta z - f(x, z, \nabla z))dx.$$

Evidently Ψ is primitive modulo $\{I\}$, and $\mathcal{E} = \{\theta, d\theta, \Psi\}$ is a Monge-Ampere system whose transverse integral manifolds (i.e., those on which $dx^1 \wedge \cdots \wedge dx^n \neq 0$) correspond to solutions of the equation (1.6). To apply our test, we start with the candidate $\Xi = \theta \wedge \Psi$, for which

$$d\Xi = -\theta \wedge d\Psi = \theta \wedge df \wedge dx.$$

Therefore, we consider φ satisfying

$$\theta \wedge df \wedge dx = \varphi \wedge \Xi,$$

or equivalently

$$df \wedge dx \equiv -\varphi \wedge \Psi \quad (\mathrm{mod}\ \{I\}),$$

and find that they are exactly those 1-forms of the form

$$\varphi = \sum f_{p_i}\, dx^i + c\theta$$

for an arbitrary function c. The problem is reduced to describing those $f(x, z, p)$ for which there exists some $c(x, z, p)$ so that $\varphi = \sum f_{p_i}\, dx^i + c\theta$ is closed. We can determine all such forms explicitly, as follows. The condition that φ be closed expands to

$$
\begin{aligned}
0 = \;& c_{p_i}\, dp_i \wedge dz \\
& + (f_{p_i p_j} - c\delta_i^j - c_{p_j} p_i)\, dp_j \wedge dx^i \\
& + \tfrac{1}{2}(f_{p_i x^j} - f_{p_j x^i} - c_{x^j} p_i + c_{x^i} p_j)\, dx^j \wedge dx^i \\
& + (f_{p_i z} - c_{x^i} - c_z p_i)\, dz \wedge dx^i.
\end{aligned}
$$

These four terms must vanish separately. The vanishing of the first term implies that $c = c(x^i, z)$ does not depend on any p_i. Given this, the vanishing of the second term implies that $f(x^i, z, p_i)$ is quadratic in the p_i, with diagonal leading term

$$f(x^i, z, p_i) = \tfrac{1}{2} c(x^i, z) \sum p_j^2 + \sum e^j(x^i, z) p_j + a(x^i, z)$$

for some functions $e^j(x^i, z)$ and $a(x^i, z)$. Now the vanishing of the third term reduces to

$$0 = e_{x^j}^i - e_{x^i}^j,$$

implying that for some function $b(x^j, z)$,

$$e^j(x^i, z) = \frac{\partial b(x^i, z)}{\partial x^j};$$

this $b(x^j, z)$ is uniquely determined only up to addition of a function of z. Finally, the vanishing of the fourth term reduces to

$$(b_z - c)_{x^i} = 0,$$

so that $c(x^i, z)$ differs from $b_z(x^i, z)$ by a function of z alone. By adding an anti-derivative of this difference to $b(x^i, z)$ and relabeling the result as $b(x^i, z)$, we see that our criterion for the Monge-Ampere system to be Euler-Lagrange is that $f(x^i, z, p_i)$ be of the form

$$f(x^i, z, p_i) = \tfrac{1}{2} b_z(x, z) \sum p_i^2 + \sum b_{x^i}(x, z) p_i + a(x, z)$$

for some functions $b(x, z)$, $a(x, z)$. These describe exactly those Poisson equations that are locally contact-equivalent to Euler-Lagrange equations.

Example 2. An example that is not quasi-linear is given by

$$\det(\nabla^2 z) - g(x, z, \nabla z) = 0.$$

The n-form $\Psi = dp - g(x, z, p)dx$ and the standard contact system generate a Monge-Ampere system whose transverse integral manifolds correspond to solutions of this equation. A calculation similar to that in the preceding example shows that this Monge-Ampere system is Euler-Lagrange if and only if $g(x, z, p)$ is of the form

$$g(x, z, p) = g_0(x, z)\, g_1(p, z - \sum p_i x^i).$$

Example 3. The linear Weingarten equation $aK + bH + c = 0$ for a surface in Euclidean space having Gauss curvature K and mean curvature H is Euler-Lagrange for all choices of constants a, b, c, as we shall see in §1.4.2. In this case, the appropriate contact manifold for the problem is $M = G_2(T\mathbf{E}^3)$, the Grassmannian of oriented tangent planes of Euclidean space.

Example 4. Here is an example of a Monge-Ampere system which is locally, but not globally, Euler-Lagrange, suitable for those readers familiar with some complex algebraic geometry. Let X be a K3 surface; that is, X is a simply connected, compact, complex manifold of complex dimension 2 with trivial canonical bundle, necessarily of Kähler type. Suppose also that there is a positive holomorphic line bundle $L \to X$ with a Hermitian metric having positive first Chern form $\omega \in \Omega^{1,1}(X)$. Our contact manifold M is the unit circle subbundle of $L \to X$, a smooth manifold of real dimension 5; the contact form is

$$\theta = \tfrac{i}{2\pi}\alpha, \qquad d\theta = \omega,$$

where α is the $\mathfrak{u}(1)$-valued Hermitian connection form on M. Note $\theta \wedge (d\theta)^2 \neq 0$, because the 4-form $(d\theta)^2 = \omega^2$ is actually a volume form on M (by positivity) and θ is non-vanishing on fibers of $M \to X$, unlike $(d\theta)^2$.

Now we trivialize the canonical bundle of X with a holomorphic 2-form $\Phi = \Psi + i\Sigma$, and take for our Monge-Ampere system

$$\mathcal{E} = \{\theta, d\theta = \omega, \Psi = \operatorname{Re}(\Phi)\}.$$

We can see that $\mathcal{E}$ is locally Euler-Lagrange as follows. First, by reasons of type, $\omega \wedge \Phi = 0$; and ω is real, so $0 = \operatorname{Re}(\omega \wedge \Phi) = \omega \wedge \Psi$. In particular, Ψ is primitive. With $\Xi = \theta \wedge \Psi$, we compute

$$d\Xi = \omega \wedge \Psi - \theta \wedge d\Psi = -\theta \wedge d\Psi,$$

but $d\Psi = \mathrm{Re}(d\Phi) = 0$, because Φ is holomorphic and therefore closed.

On the other hand, $(M, \mathcal{E})$ cannot be globally Euler-Lagrange; that is, $\Xi = \theta \wedge \Psi$ cannot be exact, for if $\Xi = d\xi$, then

$$\int_M \Xi \wedge \Sigma = \int_M d(\xi \wedge \Sigma) = 0,$$

but also

$$\int_M \Xi \wedge \Sigma = \int_M \theta \wedge \Psi \wedge \Sigma = c \int_X \Phi \wedge \bar{\Phi},$$

for some number $c \neq 0$.

1.3   Noether's Theorem

The classical theorem of Noether describes an isomorphism between a Lie algebra of infinitesimal symmetries associated to a variational problem and a space of conservation laws for its Euler-Lagrange equations. We will often assume without comment that our Lagrangian is non-degenerate in the sense discussed earlier.

There are four reasonable Lie algebras of symmetries that we might consider in our setup. Letting $\mathcal{V}(M)$ denote the Lie algebra of all vector fields on M, they are the following.

- Symmetries of (M, I, Λ):

$$\mathfrak{g}_\Lambda = \{v \in \mathcal{V}(M) : \mathcal{L}_v I \subseteq I, \ \mathcal{L}_v \Lambda = 0\}.$$

- Symmetries of $(M, I, [\Lambda])$:

$$\mathfrak{g}_{[\Lambda]} = \{v \in \mathcal{V}(M) : \mathcal{L}_v I \subseteq I, \ \mathcal{L}_v [\Lambda] = 0\}.$$

- Symmetries of (M, Π):

$$\mathfrak{g}_\Pi = \{v \in \mathcal{V}(M) : \mathcal{L}_v \Pi = 0\}.$$

(Note that $\mathcal{L}_v \Pi = 0$ implies $\mathcal{L}_v I \subseteq I$ for non-degenerate Λ.)

- Symmetries of $(M, \mathcal{E}_\Lambda)$:

$$\mathfrak{g}_{\mathcal{E}_\Lambda} = \{v \in \mathcal{V}(M) : \mathcal{L}_v \mathcal{E}_\Lambda \subseteq \mathcal{E}_\Lambda\}.$$

(Note that $\mathcal{L}_v \mathcal{E}_\Lambda \subseteq \mathcal{E}_\Lambda$ implies $\mathcal{L}_v I \subseteq I$.)

We comment on the relationship between these spaces. Clearly, there are inclusions

$$\mathfrak{g}_\Lambda \subseteq \mathfrak{g}_{[\Lambda]} \subseteq \mathfrak{g}_\Pi \subseteq \mathfrak{g}_{\mathcal{E}_\Lambda}.$$

Any of the three inclusions may be strict. For example, we *locally* have $\mathfrak{g}_{[\Lambda]} = \mathfrak{g}_\Pi$ because Π is the image of $[\Lambda]$ under the coboundary $\delta : H^n(\Omega^*/\mathcal{I}) \to H^{n+1}(\mathcal{I})$, which is invariant under diffeomorphisms of (M, I) and is an isomorphism on contractible open sets. However, we shall see later that globally there is an inclusion

$$\mathfrak{g}_\Pi / \mathfrak{g}_{[\Lambda]} \hookrightarrow H^n_{dR}(M),$$

and this discrepancy between the two symmetry algebras introduces some subtlety into Noether's theorem.

Also, there is a bound

$$\dim \left(\mathfrak{g}_{\mathcal{E}_\Lambda} / \mathfrak{g}_\Pi \right) \leq 1. \tag{1.7}$$

This follows from noting that if a vector field v preserves $\mathcal{E}_\Lambda$, then it preserves Π up to multiplication by a function; that is, $\mathcal{L}_v \Pi = f\Pi$. Because Π is a closed form, we find that $df \wedge \Pi = 0$; in the non-degenerate case, this implies $df = u\theta$ for some function u. The definition of a contact form prohibits any $u\theta$ from being closed unless $u = 0$, meaning that f is a constant. This constant gives a linear functional on $\mathfrak{g}_{\mathcal{E}_\Lambda}$ whose kernel is $\mathfrak{g}_\Pi$, proving (1.7). The area functional and minimal surface equation for Euclidean hypersurfaces provide an example where the two spaces are different. In that case, the induced Monge-Ampere system is invariant not only under Euclidean motions, but under dilations of Euclidean space as well; this is not true of the Poincaré-Cartan form.

The next step in introducing Noether's theorem is to describe the relevant spaces of conservation laws. In general, suppose that $(M, \mathcal{J})$ is an exterior differential system with integral manifolds of dimension n. A *conservation law* for $(M, \mathcal{J})$ is an $(n-1)$-form $\varphi \in \Omega^{n-1}(M)$ such that $d(f^*\varphi) = 0$ for every integral manifold $f : N^n \hookrightarrow M$ of $\mathcal{J}$. Actually, we will only consider as conservation laws those φ on M such that $d\varphi \in \mathcal{J}$, which may be a strictly smaller set. This will not present any liability, as one can always "saturate" $\mathcal{J}$ to remove this discrepancy. The two apparent ways in which a conservation law may be *trivial* are when either $\varphi \in \mathcal{J}^{n-1}$ already or φ is exact on M. Factoring out these cases leads us to the following.

Definition 1.5 *The space of conservation laws for $(M, \mathcal{J})$ is*

$$\mathcal{C} = H^{n-1}(\Omega^*(M)/\mathcal{J}).$$

It also makes sense to factor out those conservation laws represented by $\varphi \in \Omega^{n-1}(M)$ which are already closed on M, and not merely on integral manifolds of $\mathcal{J}$. This can be understood using the long exact sequence:

$$\cdots \to H^{n-1}_{dR}(M) \xrightarrow{\pi} \mathcal{C} \to H^n(\mathcal{J}) \to H^n_{dR}(M) \to \cdots.$$

Definition 1.6 *The space of proper conservation laws is* $\bar{\mathcal{C}} = \mathcal{C}/\pi(H^{n-1}_{dR}(M))$.

Note that there is an inclusion $\bar{\mathcal{C}} \hookrightarrow H^n(\mathcal{J})$. In case $\mathcal{J} = \mathcal{E}_\Lambda$ is the Euler-Lagrange system of a non-degenerate functional Λ on a contact manifold (M, I), we have the following.

Theorem 1.3 (Noether) *Let $(M, \mathcal{E}_\Lambda)$ be the Euler-Lagrange system of a non-degenerate functional Λ. There is a linear isomorphism*

$$\eta : \mathfrak{g}_{\Pi} \to H^n(\mathcal{E}_\Lambda),$$

taking the subalgebra $\mathfrak{g}_{[\Lambda]} \subset \mathfrak{g}_{\Pi}$ to the subspace $\eta(\mathfrak{g}_{[\Lambda]}) = \bar{\mathcal{C}} \subset H^n(\mathcal{E}_\Lambda)$.

Before proceeding to the proof, which will furnish an explicit formula for η, we need to make a digression on the algebra of infinitesimal contact transformations

$$\mathfrak{g}_I = \{v \in \mathcal{V}(M) : \mathcal{L}_v I \subseteq I\}.$$

The key facts are that on any neighborhood where I has a non-zero generator θ, a contact symmetry v is uniquely determined by its so-called *generating function* $g = v \lrcorner \theta$, and that given such θ, any function g is the generating function of some $v \in \mathfrak{g}_I$. This can be seen on a possibly smaller neighborhood by taking Pfaff coordinates with $\theta = dz - \sum p_i dx^i$. Working in a basis $\partial_\theta, \partial^i, \partial_i$ dual to the basis θ, dp_i, dx^i of T^*M, we write

$$v = g\,\partial_\theta + \sum v^i \partial_i + \sum v_i \partial^i.$$

Now the condition

$$\mathcal{L}_v \theta \equiv 0 \pmod{\{I\}}$$

can be made explicit, and it turns out to be

$$v_i = \partial_i g = \left(\frac{\partial}{\partial x^i} + p_i \frac{\partial}{\partial z}\right) g, \quad v^i = -\partial^i g = -\frac{\partial g}{\partial p_i}.$$

This establishes our claim, because the correspondence between v and g is now given by

$$v = g\partial_\theta - \sum g_{p_i} \partial_i + \sum (g_{x^i} + p_i g_z)\partial^i. \tag{1.8}$$

As we have presented it, the correspondence between infinitesimal contact symmetries and their generating functions is local. But a simple patching argument shows that globally, as one moves between different local generators θ for I, the different generating functions g glue together to give a global section $g \in \Gamma(M, I^*)$ of the dual line bundle. In fact, the formula (1.8) describes a canonical splitting of the surjection

$$\Gamma(TM) \to \Gamma(I^*) \to 0.$$

Note that this splitting is not a bundle map, but a differential operator.

Returning to Noether's theorem, the proof that we present is slightly incomplete in that we assume given a global non-zero contact form $\theta \in \Gamma(I)$, or equivalently, that the contact line bundle is trivial. This allows us to treat generating functions of contact symmetries as functions rather than as sections of I^*. It is an enlightening exercise to develop the patching arguments needed to overcome this using sheaf cohomology. Alternatively, one can simply pull everything up to a double cover of M on which I has a global generator, and little will be lost.

Proof of Theorem 1.3. *Step 1: Definition of the map η.* The map in question is given by

$$\eta(v) = v \lrcorner \Pi \qquad \text{for } v \in \mathfrak{g}_\Pi \subset \mathcal{V}(M).$$

Note that locally $v \lrcorner \Pi = (v \lrcorner \theta)\Psi - \theta \wedge (v \lrcorner \Psi)$, so that $v \lrcorner \Pi$ lies in $\mathcal{E}_\Lambda$. Furthermore, the condition $\mathcal{L}_v \Pi = 0$ gives

$$0 = v \lrcorner d\Pi + d(v \lrcorner \Pi) = d(v \lrcorner \Pi),$$

so that $v \lrcorner \Pi$ is closed, and gives a well-defined class $\eta(v) \in H^n(\mathcal{E}_\Lambda)$.

Step 2: η is injective. Write

$$\eta(v) = (v \lrcorner \theta)\Psi - \theta \wedge (v \lrcorner \Psi).$$

Suppose that this n-form is cohomologous to zero in $H^n(\mathcal{E}_\Lambda)$; that is,

$$\begin{aligned}
(v \lrcorner \theta)\Psi - \theta \wedge (v \lrcorner \Psi) &= d(\theta \wedge \alpha + d\theta \wedge \beta) \\
&= -\theta \wedge d\alpha + d\theta \wedge (\alpha + d\beta).
\end{aligned}$$

Regarding this equation modulo $\{I\}$ and using the primitivity of Ψ, we conclude that

$$v \lrcorner \theta = 0.$$

An infinitesimal symmetry $v \in \mathfrak{g}_I$ of the contact system is locally determined by its generating function $v \lrcorner \theta$ as in (1.8), so we conclude that $v = 0$, proving injectivity.

Step 3: η is locally surjective. We start by representing a class in $H^n(\mathcal{E}_\Lambda)$ by a closed n-form

$$\Phi = g\Psi + \theta \wedge \alpha. \tag{1.9}$$

We can choose the unique contact vector field v such that $v \lrcorner \theta = g$, and our goals are to show that $v \in \mathfrak{g}_\Pi$ and that $\eta(v) = [\Phi] \in H^n(\mathcal{E}_\Lambda)$.

For this, we need a special choice of Ψ, which so far is determined only modulo $\{I\}$; this is reasonable because the presentation (1.9) is not unique. In fact, we can further normalize Ψ by the condition

$$d\theta \wedge \Psi = 0.$$

To see why this is so, first note that by symplectic linear algebra (Proposition 1.1),

$$d\Psi \equiv d\theta \wedge \Gamma \pmod{\{I\}}, \tag{1.10}$$

for some Γ, because $d\Psi$ is of degree $n + 1$. Now suppose we replace Ψ by $\bar{\Psi} = \Psi - \theta \wedge \Gamma$, which certainly preserves the essential condition $\Pi = \theta \wedge \bar{\Psi}$. Then we have

$$
\begin{aligned}
d\theta \wedge \bar{\Psi} &= d\theta \wedge (\Psi - \theta \wedge \Gamma) \\
&= (d\Pi + \theta \wedge d\Psi) - d\theta \wedge \theta \wedge \Gamma \\
&= \theta \wedge (d\Psi - d\theta \wedge \Gamma) \\
&= 0, \text{ by } (1.10),
\end{aligned}
$$

and we have obtained our refined normalization.

Now we combine the following three equations modulo $\{I\}$:

- $0 \equiv \mathcal{L}_v \theta \equiv dg + v \lrcorner d\theta$, when multiplied by Ψ, gives

$$
dg \wedge \Psi + (v \lrcorner d\theta) \wedge \Psi \equiv 0 \quad (\text{mod } \{I\});
$$

- $0 = d\Phi = d(g\Psi + \theta \wedge \alpha)$, so using our normalization condition $d\theta \wedge \Psi = 0$ (which implies $d\Psi \equiv 0 \pmod{\{I\}}$),

$$
dg \wedge \Psi + d\theta \wedge \alpha \equiv 0 \quad (\text{mod } \{I\});
$$

- Ordinary primitivity gives $d\theta \wedge \Psi \equiv 0 \pmod{\{I\}}$, and contracting with v,

$$
(v \lrcorner d\theta) \wedge \Psi + d\theta \wedge (v \lrcorner \Psi) \equiv 0 \quad (\text{mod } \{I\}).
$$

These three equations combine to give

$$
d\theta \wedge (\alpha + v \lrcorner \Psi) \equiv 0 \quad (\text{mod } \{I\}),
$$

and from symplectic linear algebra, we have

$$
\alpha + v \lrcorner \Psi \equiv 0 \quad (\text{mod } \{I\}).
$$

This allows us to conclude

$$
v \lrcorner \Pi = g\Psi + \theta \wedge \alpha, \tag{1.11}
$$

which would complete the proof of surjectivity, except that we have not yet shown that $\mathcal{L}_v \Pi = 0$. However, by hypothesis $g\Psi + \theta \wedge \alpha$ is closed; with (1.11), this is enough to compute $\mathcal{L}_v \Pi = 0$.

The global isomorphism asserted in the theorem follows easily from these local conclusions, so long as we maintain the assumption that there exists a global contact form.

Step 4: η maps symmetries of $[\Lambda]$ to proper conservation laws. For this, first note that there is an exact sequence

$$
0 \to \bar{C} \to H^n(\mathcal{E}_\Lambda) \xrightarrow{i} H^n_{dR}(M) \to \cdots
$$

so it suffices to show that for $v \in \mathfrak{g}_\Pi$,

$$\mathcal{L}_v[\Lambda] = 0 \text{ if and only if } \eta(v) \in \text{Ker } i. \tag{1.12}$$

Recall that $\Pi = d(\Lambda - \theta \wedge \beta)$ for some β, and we can therefore calculate

$$
\begin{aligned}
\eta(v) &= v \lrcorner \Pi \\
&= \mathcal{L}_v(\Lambda - \theta \wedge \beta) - d(v \lrcorner (\Lambda - \theta \wedge \beta)) \\
&\equiv \mathcal{L}_v \Lambda \pmod{d\Omega^{n-1}(M) + \mathcal{I}^n}.
\end{aligned}
$$

This proves that $j \circ i(\eta(v)) = \mathcal{L}_v[\Lambda]$ in the composition

$$H^n(\mathcal{E}_\Lambda) \xrightarrow{i} H^n_{dR}(M) \xrightarrow{j} H^n(\Omega^*(M)/\mathcal{I}).$$

The conclusion (1.12) will follow if we can prove that j is injective.

To see that j is injective, note that it occurs in the long exact cohomology sequence of

$$0 \to \mathcal{I} \to \Omega^*(M) \to \Omega^*(M)/\mathcal{I} \to 0;$$

namely, we have

$$\cdots \to H^n(\mathcal{I}) \to H^n_{dR}(M) \xrightarrow{j} H^n(\Omega^*(M)/\mathcal{I}) \to \cdots.$$

So it suffices to show that $H^n(\mathcal{I}) = 0$, which we will do under the standing assumption that there is a global contact form θ. Suppose that the n-form

$$\varphi = \theta \wedge \alpha + d\theta \wedge \beta = \theta \wedge (\alpha + d\beta) + d(\theta \wedge \beta)$$

is closed. Then regarding $0 = d\varphi$ modulo $\{I\}$, we have by symplectic linear algebra that

$$\alpha + d\beta \equiv 0 \pmod{\{I\}}$$

so that actually

$$\varphi = d(\theta \wedge \beta).$$

This says that $\varphi \sim 0$ in $H^n(\mathcal{I})$, and our proof is complete. $\qquad\square$

It is important in practice to have a local formula for a representative in $\Omega^{n-1}(M)$, closed modulo $\mathcal{E}_\Lambda$, for the proper conservation law $\eta(v)$. This is obtained by first writing as usual

$$\Pi = d(\Lambda - \theta \wedge \beta), \tag{1.13}$$

and also, for a given $v \in \mathfrak{g}_{[\Lambda]}$,

$$\mathcal{L}_v \Lambda \equiv d\gamma \pmod{\mathcal{I}}. \tag{1.14}$$

We will show that the $(n-1)$-form

$$\boxed{\varphi = -v \lrcorner \Lambda + (v \lrcorner \theta)\beta + \gamma} \tag{1.15}$$

is satisfactory. First, compute

$$
\begin{aligned}
d\varphi &= (-\mathcal{L}_v\Lambda + v \lrcorner \, d\Lambda) + d((v \lrcorner \, \theta)\beta) + d\gamma \\
&\equiv v \lrcorner \, (\Pi + d(\theta \wedge \beta)) + d((v \lrcorner \, \theta)\beta) \quad (\mathrm{mod} \ \mathcal{I}) \\
&\equiv \eta(v) + \mathcal{L}_v(\theta \wedge \beta) \quad (\mathrm{mod} \ \mathcal{I}) \\
&\equiv \eta(v) \quad (\mathrm{mod} \ \mathcal{I}).
\end{aligned}
$$

Now we have $d\varphi = \eta(v) + \Xi$ for some closed $\Xi \in \mathcal{I}^n$. We proved in the last part of the proof of Noether's theorem that $H^n(\mathcal{I}) = 0$, which implies that $\Xi = d\xi$ for some $\xi \in \mathcal{I}^{n-1}$. Now we have $d(\varphi - \xi) = \eta(v)$, and $\varphi \sim \varphi - \xi$ in $\mathcal{C} = H^{n-1}(\Omega^*(M)/\mathcal{E}_\Lambda)$. This justifies our prescription (1.15).

Note that the prescription is especially simple when $v \in \mathfrak{g}_\Lambda \subseteq \mathfrak{g}_{[\Lambda]}$, for then we can take $\gamma = 0$.

Example. Let $\mathbf{L}^{n+1} = \{(t, y^1, \ldots, y^n)\} \cong \mathbf{R}^{n+1}$ be Minkowski space, and let $M^{2n+3} = J^1(\mathbf{L}^{n+1}, \mathbf{R})$ be the standard contact manifold, with coordinates (t, y^i, z, p_a) (where $0 \leq a \leq n$), $\theta = dz - p_0 dt - \sum p_i dy^i$. For a Lagrangian, take

$$
\Lambda = \left(\tfrac{1}{2}\|p\|^2 + F(z) \right) dt \wedge dy
$$

for some "potential" function $F(z)$, where $dy = dy^1 \wedge \cdots \wedge dy^n$ and $\|p\|^2 = -p_0^2 + \sum p_i^2$ is the Lorentz-signature norm. The local symmetry group of this functional is generated by two subgroups, the translations in $\mathbf{L}^{n+1}$ and the linear isometries $SO^o(1, n)$; as we shall see in Chapter 3, for certain $F(z)$ the symmetry group of the associated Poincaré-Cartan form is strictly larger. For now, we calculate the conservation law corresponding to translation in t, and begin by finding the Poincaré-Cartan form Π. Letting $f(z) = F'(z)$, we differentiate

$$
\begin{aligned}
d\Lambda &= (-p_0 dp_0 + \sum p_i dp_i + f(z)\theta) \wedge dt \wedge dy \\
&= \theta \wedge (f(z)dt \wedge dy) + d\theta \wedge (p_0 dy + \sum p_j dt \wedge dy_{(j)});
\end{aligned}
$$

with the usual recipe $\Pi = \theta \wedge (\alpha + d\beta)$ whenever $d\Lambda = \theta \wedge \alpha + d\theta \wedge \beta$, we obtain

$$
\Pi = \theta \wedge \left(f(z)dt \wedge dy + dp_0 \wedge dy + \sum dp_j \wedge dt \wedge dy_{(j)} \right).
$$

We see the Euler-Lagrange equation using

$$
\Psi = f(z)dt \wedge dy + dp_0 \wedge dy + \sum dp_j \wedge dt \wedge dy_{(j)};
$$

an integral manifold of $\mathcal{E}_\Lambda = \{\theta, d\theta, \Psi\}$ of the form

$$
\{(t, y, z(t, y), z_t(t, y), z_{y^i}(t, y))\}
$$

must satisfy

$$
0 = \Psi|_N = \left(\frac{\partial^2 z}{\partial t^2} - \sum \frac{\partial^2 z}{\partial y^{i2}} + f(z) \right) dt \wedge dy.
$$

With the independence condition $dt \wedge dy \neq 0$, we have the familiar wave equation

$$
\Box z(t, y) = f(z).
$$

Now considering the time-translation symmetry $v = \frac{\partial}{\partial t} \in \mathfrak{g}_\Lambda$, the Noether prescription (1.15) gives

$$
\begin{aligned}
\varphi &= -v \lrcorner \Lambda + (v \lrcorner \theta)\beta \\
&= -\left(\frac{\|p\|^2}{2} + F(z)\right) dy - p_0 \left(p_0 dy + \sum p_j dt \wedge dy_{(j)}\right) \\
&= -\left(\tfrac{1}{2} \sum p_a^2 + F(z)\right) dy - p_0 dt \wedge \left(\sum p_j dy_{(j)}\right).
\end{aligned}
$$

One can verify that φ is closed when restricted to a solution of $\Box z = f(z)$. The question of how one might use this conservation law will be taken up later.

1.4   Hypersurfaces in Euclidean Space

We will apply the the theory developed so far to the study of hypersurfaces in Euclidean space

$$N^n \hookrightarrow \mathbf{E}^{n+1}.$$

We are particularly interested in the study of those functionals on such hypersurfaces which are invariant under the group $E(n+1)$ of orientation-preserving Euclidean motions.

1.4.1   The Contact Manifold over $\mathbf{E}^{n+1}$

Points of $\mathbf{E}^{n+1}$ will be denoted $x = (x^0, \ldots, x^n)$, and each tangent space $T_x\mathbf{E}^{n+1}$ will be canonically identified with $\mathbf{E}^{n+1}$ itself via translation. A *frame* for $\mathbf{E}^{n+1}$ is a pair

$$f = (x, e)$$

consisting of a point $x \in \mathbf{E}^{n+1}$ and a positively oriented orthonormal basis $e = (e_0, \ldots, e_n)$ for $T_x\mathbf{E}^{n+1}$. The set $\mathcal{F}$ of all such frames is a manifold, and the right $SO(n+1, \mathbf{R})$-action

$$(x, (e_0, \ldots, e_n)) \cdot (g_b^a) = (x, (\sum e_a g_0^a, \ldots, \sum e_a g_n^a))$$

gives the basepoint map

$$x : \mathcal{F} \to \mathbf{E}^{n+1}$$

the structure of a principal bundle.[5] There is also an obvious left-action of $E(n+1, \mathbf{R})$ on $\mathcal{F}$, and a choice of reference frame gives a left-equivariant identification $\mathcal{F} \cong E(n+1)$ of the bundle of frames with the group of Euclidean motions.

   The relevant contact manifold for studying hypersurfaces in $\mathbf{E}^{n+1}$ is the manifold of *contact elements*

$$M^{2n+1} = \{(x, H) : x \in \mathbf{E}^{n+1},\ H^n \subset T_x\mathbf{E}^{n+1} \text{ an oriented hyperplane}\}.$$

---

[5] Throughout this section, we use index ranges $1 \leq i, j \leq n$ and $0 \leq a, b \leq n$.

This M will be given the structure of a contact manifold in such a way that transverse Legendre submanifolds correspond to arbitrary immersed hypersurfaces in $\mathbf{E}^{n+1}$. Note that M may be identified with the unit sphere bundle of $\mathbf{E}^{n+1}$ by associating to a contact element (x, H) its oriented orthogonal complement (x, e_0). We will use this identification without further comment.

The projection $\mathcal{F} \to M$ taking $(x, (e_a)) \mapsto (x, e_0)$ is $E(n+1, \mathbf{R})$-equivariant (for the left-action). To describe the contact structure on M and to carry out calculations, we will actually work on $\mathcal{F}$ using the following structure equations. First, we define canonical 1-forms on $\mathcal{F}$ by differentiating the vector-valued coordinate functions $x(f), e_a(f)$ on $\mathcal{F}$, and decomposing the resulting vector-valued 1-forms at each $f \in \mathcal{F}$ with respect to the frame $e_a(f)$:

$$dx = \sum e_b \cdot \omega^b, \qquad de_a = \sum e_b \cdot \omega_a^b. \qquad (1.16)$$

Differentiating the relations $\langle e_a(f), e_b(f) \rangle = \delta_{ab}$ yields

$$\omega_b^a + \omega_a^b = 0.$$

The forms ω^a, ω_b^a satisfy no other linear-algebraic relations, giving a total of $(n+1) + \frac{1}{2}n(n+1) = \dim(\mathcal{F})$ independent 1-forms. By taking the derivatives of the defining relations (1.16), we obtain the structure equations

$$d\omega^a + \sum \omega_c^a \wedge \omega^c = 0, \qquad d\omega_b^a + \sum \omega_c^a \wedge \omega_b^c = 0. \qquad (1.17)$$

The forms ω^a are identified with the usual tautological 1-forms on the orthonormal frame bundle of a Riemannian manifold (in this case, of $\mathbf{E}^{n+1}$); and then the first equation indicates that ω_b^a are components of the Levi-Civita connection of $\mathbf{E}^{n+1}$, while the second indicates that it has vanishing Riemann curvature tensor.

In terms of these forms, the fibers of $x : \mathcal{F} \to \mathbf{E}^{n+1}$ are exactly the maximal connected integral manifolds of the Pfaffian system $\{\omega^a\}$. Note that $\{\omega^a\}$ and $\{dx^a\}$ are alternative bases for the space of forms on $\mathcal{F}$ that are semibasic over $\mathbf{E}^{n+1}$, but the former is $E(n+1)$-invariant, while the latter is not.

We return to an explanation of our contact manifold M, by first distinguishing the 1-form on $\mathcal{F}$

$$\theta \overset{def}{=} \omega^0.$$

Note that its defining formula

$$\theta_f(v) = \langle dx(v), e_0(f) \rangle, \qquad v \in T_f \mathcal{F},$$

shows that it is the pullback of a unique, globally defined 1-form on M, which we will also call $\theta \in \Omega^1(M)$. To see that θ is a contact form, first relabel the forms on $\mathcal{F}$ (this will be useful later, as well)

$$\pi_i \overset{def}{=} \omega_i^0,$$

and note the equation on $\mathcal{F}$

$$d\theta = -\sum \pi_i \wedge \omega^i.$$

So on $\mathcal{F}$ we certainly have $\theta \wedge (d\theta)^n \neq 0$, and because pullback of forms via the submersion $\mathcal{F} \to M$ is injective, the same non-degeneracy holds on M.

To understand the Legendre submanifolds of M, consider an oriented immersion

$$N^n \overset{\iota}{\hookrightarrow} M^{2n+1}, \quad y = (y^1, \ldots, y^n) \mapsto (x(y), e_0(y)).$$

The Legendre condition is

$$(\iota^*\theta)_y(v) = \langle dx_y(v), e_0(y) \rangle = 0, \qquad v \in T_y N.$$

In the transverse case, when the composition $x \circ \iota : N^n \hookrightarrow M^{2n+1} \to \mathbf{E}^{n+1}$ is a hypersurface immersion (equivalently, $\iota^*(\bigwedge \omega^i) \neq 0$, suitably interpreted), this condition is that $e_0(y)$ is a unit normal vector to the hypersurface $x \circ \iota(N)$. These Legendre submanifolds may therefore be thought of as the graphs of Gauss maps of oriented hypersurfaces $N^n \hookrightarrow \mathbf{E}^{n+1}$. Non-transverse Legendre submanifolds of M are sometimes of interest. To give some intuition for these, we exhibit two examples in the contact manifold over $\mathbf{E}^3$. First, over an immersed curve $x : I \hookrightarrow \mathbf{E}^3$, one can define a cylinder $N = S^1 \times I \hookrightarrow M \cong \mathbf{E}^3 \times S^2$ by

$$(v, w) \mapsto (x(w), R_v(\nu_x)),$$

where ν is any normal vector field along the curve $x(w)$, and R_v is rotation through angle $v \in S^1$ about the tangent $x'(w)$. The image is just the unit normal bundle of the curve, and it is easily verified that this is a Legendre submanifold.

Our second example corresponds to the *pseudosphere*, a singular surface of revolution in $\mathbf{E}^3$ having constant Gauss curvature $K = -1$ away from the singular locus. The map $x : S^1 \times \mathbf{R} \to \mathbf{E}^3$ given by

$$x : (v, w) \mapsto (\operatorname{sech} w \cos v, -\operatorname{sech} w \sin v, w - \tanh w)$$

fails to be an immersion where $w = 0$. However, the Gauss map of the complement of this singular locus can be extended to a smooth map $e_3 : S^1 \times \mathbf{R} \to S^2$ given by

$$e_3(v, w) = (-\tanh w \cos v, \tanh w \sin v, -\operatorname{sech} w).$$

The graph of the Gauss map is the product $(x, e_3) : S^1 \times \mathbf{R} \hookrightarrow M$. It is a Legendre submanifold, giving a smooth surface in M whose projection to $\mathbf{E}^3$ is one-to-one, is an immersion almost everywhere, and has image equal to the singular pseudosphere. We will discuss in §4.3.3 the exterior differential system whose integral manifolds are graphs of Gauss maps of $K = -1$ surfaces in $\mathbf{E}^3$. In §4.3.4, we will discuss the Bäcklund transformation for this system, which relates this particular example to a special case of the preceding example, the unit normal bundle of a line.

1.4.2   Euclidean-invariant Euler-Lagrange Systems

We can now introduce one of the most important of all variational problems, that of finding minimal-area hypersurfaces in Euclidean space. Define the n-form

$$\Lambda = \omega^1 \wedge \cdots \wedge \omega^n \in \Omega^n(\mathcal{F}),$$

and observe that it is basic over M; that is, it is the pullback of a well-defined n-form on M (although its factors ω^i are not basic). This defines a Lagrangian functional

$$\mathcal{F}_\Lambda(N) = \int_N \Lambda$$

on compact Legendre submanifolds $N^n \hookrightarrow M^{2n+1}$, which in the transverse case discussed earlier equals the area of N induced by the immersion $N \hookrightarrow \mathbf{E}^{n+1}$. We calculate the Poincaré-Cartan form up on $\mathcal{F}$ using the structure equations (1.17), as

$$d\Lambda = -\theta \wedge \sum \pi_i \wedge \omega_{(i)},$$

so the Euler-Lagrange system $\mathcal{E}_\Lambda$ is generated by $\mathcal{I} = \{\theta, d\theta\}$ and

$$\Psi = -\sum \pi_i \wedge \omega_{(i)},$$

which is again well-defined on M. A transverse Legendre submanifold $N \hookrightarrow M$ will locally have a basis of 1-forms given by pullbacks (by any section) of $\omega^1, \ldots, \omega^n$, so applying the Cartan lemma to

$$0 = d\theta|_N = -\pi_i \wedge \omega^i$$

shows that restricted to N there are expressions

$$\pi_i = \sum_j h_{ij}\omega^j$$

for some functions $h_{ij} = h_{ji}$. If $N \hookrightarrow M$ is also an integral manifold of $\mathcal{E}_\Lambda \subset \Omega^*(M)$, then additionally

$$0 = \Psi|_N = -\left(\sum h_{ii}\right) \omega^1 \wedge \cdots \wedge \omega^n.$$

One can identify h_{ij} with the second fundamental form of $N \hookrightarrow \mathbf{E}^{n+1}$ in this transverse case, and we then have the usual criterion that a hypersurface is stationary for the area functional if and only if its mean curvature $\sum h_{ii}$ vanishes. We will return to the study of this Euler-Lagrange system shortly.

Another natural $E(n+1)$-invariant PDE for hypersurfaces in Euclidean space is that of prescribed constant mean curvature H, not necessarily zero. We first ask whether such an equation is even Euler-Lagrange, and to answer this we apply our inverse problem test to the Monge-Ampère system

$$\mathcal{E}_H = \{\theta, d\theta, \Psi_H\}, \qquad \Psi_H = -\left(\sum \pi_i \wedge \omega_{(i)} - H\omega\right).$$

Here, H is the prescribed constant and $\omega = \omega^1 \wedge \cdots \wedge \omega^n$ is the induced volume form. The transverse integral manifolds of $\mathcal{E}_H$ correspond to the desired Euclidean hypersurfaces.

To implement the test, we take the candidate Poincaré-Cartan form

$$\Pi_H = -\theta \wedge \left(\sum \pi_i \wedge \omega_{(i)} - H\omega \right)$$

and differentiate; the derivative of the first term vanishes, as we know from the preceding case of $H = 0$, and we have

$$
\begin{aligned}
d\Pi_H &= H \, d(\theta \wedge \omega^1 \wedge \cdots \wedge \omega^n) \\
&= H \, d(dx^0 \wedge \cdots \wedge dx^n) \\
&= 0.
\end{aligned}
$$

So this $\mathcal{E}_H$ is at least locally the Euler-Lagrange system for some functional Λ_H, which can be taken to be an anti-derivative of Π_H. One difficulty in finding Π_H is that there is *no* such Λ_H that is invariant under the Euclidean group $E(n+1)$. The next best thing would be to find a Λ_H which is invariant under the rotation subgroup $SO(n+1, \mathbf{R})$, but not under translations. A little experimentation yields the Lagrangian

$$\Lambda_H = \omega + \tfrac{H}{n+1} x \lrcorner \Omega, \qquad d\Lambda_H = \Pi_H,$$

where $x = \sum x^a \frac{\partial}{\partial x^a}$ is the radial position vector field, $\omega = \omega^1 \wedge \cdots \wedge \omega^n$ is the hypersurface area form, and $\Omega = \omega^1 \wedge \cdots \wedge \omega^{n+1}$ is the ambient volume form. The choice of an origin from which to define the position vector x reduces the symmetry group of Λ_H from $E(n+1)$ to $SO(n+1, \mathbf{R})$. The functional $\int_N \Lambda_H$ gives the area of the hypersurface N plus a scalar multiple of the signed volume of the cone on N with vertex at the origin.

It is actually possible to list all of the Euclidean-invariant Poincaré-Cartan forms on $M \to \mathbf{E}^{n+1}$. Let

$$\Lambda_{-1} = -\tfrac{1}{n+1} x \lrcorner \Omega, \qquad \Lambda_k = \sum_{|I|=k} \pi_I \wedge \omega_{(I)} \quad (0 \le k \le n),$$

and

$$\Pi_k = -\theta \wedge \Lambda_k,$$

It is an exercise using the structure equations to show that

$$d\Lambda_k = \Pi_{k+1}.$$

Although these forms are initially defined up on $\mathcal{F}$, it is easily verified that they are pull-backs of forms on M, which we denote by the same name. It can be proved using the first fundamental theorem of orthogonal invariants that any Euclidean-invariant Poincaré-Cartan form is a linear combination of $\Pi_0, \ldots, \Pi_n$. Note that such a Poincaré-Cartan form is induced by a Euclidean-invariant functional if and only if Π_0 is not involved.

We can geometrically interpret $\Lambda_k|_N$ for transverse Legendre submanifolds N as the sum of the $k \times k$ minor determinants of the second fundamental form II_N, times the hypersurface area form of N. In case $k = n$ we have $d\Lambda_n = II_{n+1} = 0$, reflecting the fact that the functional

$$\int_N \Lambda_n = \int_N K\, dA$$

is variationally trivial, where K is the Gauss-Kronecker curvature.

Contact Equivalence of Linear Weingarten Equations for Surfaces

The Euclidean-invariant Poincaré-Cartan forms for surfaces in $\mathbf{E}^3$ give rise to the *linear Weingarten equations*, of the form

$$aK + bH + c = 0$$

for constants a, b, c. Although these second-order PDEs are inequivalent under point-transformations for non-proportional choices of a, b, c, we will show that under *contact* transformations there are only five distinct equivalence classes of linear Weingarten equations.

To study surfaces, we work on the unit sphere bundle $\pi : M^5 \to \mathbf{E}^3$, and recall the formula for the contact form

$$\theta_{(x,e_0)}(v) = \langle \pi_*(v), e_0 \rangle, \qquad v \in T_{(x,e_0)}M.$$

We define two 1-parameter groups of diffeomorphisms of M as follows:

$$\begin{aligned}
\varphi_t(x, e_0) &= (x + te_0, e_0), \\
\psi_s(x, e_0) &= (\exp(s)x, e_0).
\end{aligned}$$

It is not hard to see geometrically that these define contact transformations on M, although this result will also come out of the following calculations. We will carry out calculations on the full Euclidean frame bundle $\mathcal{F} \to \mathbf{E}^3$, where there is a basis of 1-forms ω^1, ω^2, θ, π_1, π_2, ω_2^1 satisfying structure equations presented earlier.

To study φ_t we use its generating vector field $v = \frac{\partial}{\partial\theta}$, which is the dual of the 1-form θ with respect to the preceding basis. We can easily compute Lie derivatives

$$\mathcal{L}_v\omega^1 = -\pi_1, \quad \mathcal{L}_v\omega^2 = -\pi_2, \quad \mathcal{L}_v\theta = 0, \quad \mathcal{L}_v\pi_1 = 0, \quad \mathcal{L}_v\pi_2 = 0.$$

Now, the fibers of $\mathcal{F} \to M$ have tangent spaces given by $\{\omega^1, \omega^2, \theta, \pi_1, \pi_2\}^\perp$, and this distribution is evidently preserved by the flow along v. This implies that v induces a vector field downstairs on M, whose flow is easily seen to be φ_t. The fact that $\mathcal{L}_v\theta = 0$ confirms that φ_t is a contact transformation.

We can now examine the effect of φ_t on the invariant Euler-Lagrange systems corresponding to linear Weingarten equations by introducing

$$\Psi_2 = \pi_1 \wedge \pi_2, \quad \Psi_1 = \pi_1 \wedge \omega^2 - \pi_2 \wedge \omega^1, \quad \Psi_0 = \omega^1 \wedge \omega^2.$$

Restricted to a transverse Legendre submanifold over a surface $N \subset \mathbf{E}^3$, these give $K\,dA$, $H\,dA$, and the area form dA of N, respectively. Linear Weingarten surfaces are integral manifolds of a Monge-Ampere system

$$\{\theta, d\theta, \Psi(a,b,c) \overset{def}{=} a\Psi_2 + b\Psi_1 + c\Psi_0\}.$$

Our previous Lie derivative computations may be used to compute

$$\mathcal{L}_v \begin{pmatrix} \Psi_0 \\ \Psi_1 \\ \Psi_2 \end{pmatrix} = \begin{pmatrix} 0 & -1 & 0 \\ 0 & 0 & -2 \\ 0 & 0 & 0 \end{pmatrix} \begin{pmatrix} \Psi_0 \\ \Psi_1 \\ \Psi_2 \end{pmatrix}.$$

Exponentiate this to see

$$\varphi_t^* \Psi(a,b,c) = \Psi(a - 2bt + ct^2, b - ct, c). \tag{1.18}$$

This describes how the 1-parameter group φ_t acts on the collection of linear Weingarten equations. Similar calculations show that the 1-parameter group ψ_s introduced earlier consists of contact transformations, and acts on linear Weingarten equations as

$$\psi_s^* \Psi(a,b,c) = \Psi(a, \exp(s)b, \exp(2s)c). \tag{1.19}$$

It is reasonable to regard the coefficients (a,b,c) which specify a particular linear Weingarten equation as a point $[a:b:c]$ in the real projective plane $\mathbf{RP}^2$, and it is an easy exercise to determine the orbits in $\mathbf{RP}^2$ of the group action generated by (1.18) and (1.19). There are five orbits, represented by the points $[1:0:0]$, $[0:1:0]$, $[1:0:1]$, $[1:0:-1]$, $[0:0:1]$. The special case

$$\varphi_{\frac{1}{A}}^* \Psi(0,1,A) = \Psi(-\tfrac{1}{A}, 0, A)$$

gives the classically known fact that to every surface of non-zero constant mean curvature $-A$, there is a (possibly singular) parallel surface of constant positive Gauss curvature A^2. Note finally that the Monge-Ampere system corresponding to $[0:0:1]$ has for integral manifolds those non-transverse Legendre submanifolds of M which project to curves in $\mathbf{E}^3$, instead of surfaces.

1.4.3   Conservation Laws for Minimal Hypersurfaces

In Chapter 3, we will be concerned with conservation laws for various Euler-Lagrange equations arising in conformal geometry. We will emphasize two questions: how are conservation laws found, and how can they be used? In this section, we will explore these two questions in the case of the minimal hypersurface equation $H = 0$, regarding conservation laws arising from Euclidean symmetries.

We compute these conservation laws first for the translations, and then for the rotations. The results of these computations will be the two vector-valued conservation laws

$$\boxed{d(*dx) = 0, \qquad d(*(x \wedge dx)) = 0.}$$

The notation will be explained in the course of the calculation. These may be thought of as analogs of the conservation of linear and angular momentum that are ubiquitous in physics.

To carry out the computation, note that the prescription for Noether's theorem given in (1.13, 1.14, 1.15) is particularly simple for the case of the functional

$$\Lambda = \omega^1 \wedge \cdots \wedge \omega^n$$

on the contact manifold M^{2n+1}. This is because first, $d\Lambda = \Pi$ already, so no correction term is needed, and second, the infinitesimal Euclidean symmetries (prolonged to act on M) actually preserve Λ, and not merely the equivalence class $[\Lambda]$. Consequently, the Noether prescription is (with a sign change)

$$\eta(v) = v \,\lrcorner\, \Lambda.$$

This $v \,\lrcorner\, \Lambda$ is an $(n-1)$-form on M which is closed modulo the Monge-Ampere system $\mathcal{E}_\Lambda$.

Proceeding, we can suppose that our translation vector field is written up on the Euclidean frame bundle as

$$v_{\mathcal{F}} = Ae_0 + A^i e_i,$$

where the coefficients are such that the equation $dv = 0$ holds; that is, the functions A and A^i are the coefficients of a fixed vector with respect to a varying oriented orthonormal frame. We easily find

$$\varphi_v = v_{\mathcal{F}} \,\lrcorner\, \Lambda = \sum_{i=1}^{n} A^i \omega_{(i)}.$$

This, then, is the formula for an $(n-1)$-form on $\mathcal{F}$ which is well-defined on the contact manifold M and is closed when restricted to integral manifolds of the Monge-Ampere system $\mathcal{E}_\Lambda$. To see it in another form, observe that if we restrict our $(n-1)$-form to a transverse Legendre submanifold N,

$$\varphi_v|_N = \sum A^i (*\omega^i) = *\langle v, dx \rangle.$$

Here and throughout, the star operator $* = *_N$ is defined with respect to the induced metric and orientation on N, and the last equality follows from the equation of $\mathbf{E}^{n+1}$-valued 1-forms $dx = e_0\theta + \sum e_i\omega^i$, where $\theta|_N = 0$. We now have a linear map from $\mathbf{R}^{n+1}$, regarded as the space of translation vectors v, to the space of closed $(n-1)$-forms on any minimal hypersurface N. Tautologically, such a map may be regarded as one closed $(\mathbf{R}^{n+1})^*$-valued $(n-1)$-form on N. Using the metric to identify $(\mathbf{R}^{n+1})^* \cong \mathbf{R}^{n+1}$, this may be written as

$$\varphi_{trans} = *dx.$$

This is the meaning of the conservation law stated at the beginning of this section. Note that each component $d(*dx^a) = 0$ of this conservation law is equivalent to the claim that the coordinate function x^a of the immersion x : $N \hookrightarrow \mathbf{E}^{n+1}$ is a harmonic function with respect to the induced metric on N.

Turning to the rotation vector fields, we first write such a vector field on $\mathbf{E}^{n+1}$ as

$$v = \sum_{a,b=1}^{n+1} x^a R_a^b \frac{\partial}{\partial x^b}, \qquad R_b^a + R_a^b = 0.$$

It is not hard to verify that this vector field lifts naturally to the frame bundle $\mathcal{F}$ as

$$v_{\mathcal{F}} = \sum x^a R_a^b A_c^b \frac{\partial}{\partial \omega^c} + \sum A_c^b R_a^b A_d^a \frac{\partial}{\partial \omega_d^c},$$

where the coefficients A_b^a are defined by the equation $\frac{\partial}{\partial \omega^a} = \sum A_a^b \frac{\partial}{\partial x^b}$, and the tangent vectors $\frac{\partial}{\partial \omega^a}, \frac{\partial}{\partial \omega_b^a}$ are dual to the canonical coframing ω^a, ω_b^a of $\mathcal{F}$.

We can now compute (restricted to N, for convenience)

$$\begin{aligned}(v_{\mathcal{F}} \lrcorner \Lambda)|_N &= \sum x^a R_a^b A_i^b \omega_{(i)} \\ &= *(x^a R_a^b A_c^b \omega^c) \\ &= *\langle R \cdot x, dx \rangle.\end{aligned}$$

Reformulating the Noether map in a manner analogous to that used previously, we can define a $\bigwedge^2 \mathbf{R}^{n+1} \cong \mathfrak{so}(n+1, \mathbf{R})^*$-valued $(n-1)$-form on N

$$\varphi_{rot} = *(x \wedge dx).$$

Once again, φ_{rot} is a conservation law by virtue of the fact that it is closed if N is a minimal hypersurface.

It is interesting to note that the conservation law for rotation symmetry is a consequence of that for translation symmetry. This is because we have from $d(*dx) = 0$ that

$$d(x \wedge *dx) = dx \wedge *dx = 0.$$

The last equation holds because the exterior multiplication $\wedge$ refers to the $\mathbf{E}^{n+1}$ where the forms take values, *not* the exterior algebra in which their components live. It is an exercise to show that these translation conservation laws are equivalent to minimality of N.

Another worthwhile exercise is to show that all of the classical conservation laws for the $H = 0$ system arise from infinitesimal Euclidean symmetries. In the next chapter, we will see directly that the group of symmetries of the Poincaré-Cartan form for this system equals the group of Euclidean motions, giving a more illuminating proof of this fact. At the end of this section, we will consider a dilation vector field which preserves the minimal surface system $\mathcal{E}$, but not the Poincaré-Cartan form, and use it to compute an "almost-conservation law."

By contrast, in this case there is *no* discrepancy between $\mathfrak{g}_{[\Lambda]}$ and $\mathfrak{g}_\Pi$. To see this, first note that by Noether's theorem 1.3, $\mathfrak{g}_\Pi$ is identified with $H^n(\mathcal{E}_\Lambda)$, and $\mathfrak{g}_{[\Lambda]} \subseteq \mathfrak{g}_\Pi$ is identified with the image of the connecting map δ in the long exact sequence

$$\cdots \to H^{n-1}(\Omega^*/\mathcal{E}_\Lambda) \xrightarrow{\delta} H^n(\mathcal{E}_\Lambda) \xrightarrow{\iota} H^n_{dR}(M) \to \cdots .$$

With $M \cong \mathbf{E}^{n+1} \times S^n$, we have the isomorphism $H^n_{dR}(M) \cong \mathbf{R}$ obtained by integrating an n-form along a fiber of $M \to \mathbf{E}^{n+1}$, and it is not hard to see that any n-form in $\mathcal{E}_\Lambda$ must vanish when restricted to such a fiber. Therefore the map ι is identically 0, so δ is onto, and that proves our claim.

Interpreting the Conservation Laws for $H = 0$

To understand the meaning of the conservation law φ_{trans}, we convert the equation $d\varphi_{trans}|_N = 0$ to integral form. For a smoothly bounded, oriented neighborhood $U \subset N \subset \mathbf{E}^{n+1}$ with N minimal, we have by Stokes' theorem

$$\int_{\partial U} *dx = 0.$$

To interpret this condition on U, we take an oriented orthonormal frame field $(e_0, \ldots, e_n)$ along $U \cup \partial U$, such that along the boundary ∂U the following hold:

$$\begin{cases} e_0 \text{ is the oriented normal to } N, \\ e_n \text{ is the outward normal to } \partial U \text{ in } N, \\ e_1, \ldots, e_{n-1} \text{ are tangent to } \partial U. \end{cases} \qquad (1.20)$$

Calculations will be much easier in this adapted frame field. The dual coframe ω^a for $\mathbf{E}^{n+1}$ along $U \cup \partial U$ satisfies

$$dx = e_0\omega^0 + \sum_{i=1}^{n-1} e_i\omega^i + e_n\omega^n.$$

Now, the first term vanishes when restricted to N. The last term vanishes when restricted to ∂U, but cannot be discarded because it will affect $*_N dx$, which we are trying to compute. Consequently,

$$*_N dx = \sum_{i=1}^{n-1} e_i\omega_{(i)} + e_n\omega_{(n)}.$$

Now we restrict to ∂U, and find

$$\begin{aligned} *_N dx|_{\partial U} &= (-1)^{n-1}e_n\omega^1 \wedge \cdots \wedge \omega^{n-1} \\ &= (-1)^{n-1}\mathbf{n}\, d\sigma. \end{aligned}$$

Here we use $\mathbf{n}$ to denote the normal to ∂U in N and $d\sigma$ to denote the area measure induced on ∂U. Our conservation law therefore reads

$$\int_{\partial U} \mathbf{n}\, d\sigma = 0.$$

In other words, in a minimal hypersurface the average of the exterior unit normal vectors over the smooth boundary of any oriented neighborhood must vanish. One consequence of this is that a minimal surface can never be locally convex; that is, a neighborhood of a point can never lie on one side of the tangent plane at that point. This is intuitively reasonable from the notion of minimality. Similar calculations give an analogous formulation for the rotation conservation law:

$$\int_{\partial U} (\mathbf{x} \wedge \mathbf{n}) d\sigma = 0.$$

These interpretations have relevance to the classical Plateau problem, which asks whether a given simple closed curve γ in $\mathbf{E}^3$ bounds a minimal surface. The answer to this is affirmative, with the caveat that such a surface is not necessarily unique and may not be smooth at the boundary. A more well-posed version gives not only a simple closed curve $\gamma \subset \mathbf{E}^3$, but a *strip*, which is a curve $\gamma^{(1)} \subset M$ consisting of a base curve $\gamma \subset \mathbf{E}^3$ along with a field of tangent planes along γ containing the tangent lines to γ. Such a strip is the same as a curve in M along which the contact 1-form vanishes. Asking for a minimal surface whose boundary and boundary-tangent planes are described by a given $\gamma^{(1)}$ is the same as asking for a transverse integral manifold of $\mathcal{E}_\Lambda$ having boundary $\gamma^{(1)} \subset M$.

. The use of our two conservation laws in this context comes from the fact that $\gamma^{(1)}$ determines the vector-valued form $*_N dx$ along ∂N for *any* possible solution to this initial value problem. The conservation laws give integral constraints, often called *moment conditions*, on the values of $*_N dx$, and hence constrain the possible strips $\gamma^{(1)}$ for which our problem has an affirmative answer. However, the moment conditions on a strip $\gamma^{(1)}$ are not sufficient for there to exist a minimal surface with that boundary data. We will discuss additional constraints which have the feel of "hidden conservation laws" after a digression on similar moment conditions that arise for boundaries of holomorphic curves.

It is natural to ask whether a given real, simple, closed curve $\gamma_{\mathbf{C}}$ in complex space $\mathbf{C}^n$ (always $n \geq 2$) is the boundary of some holomorphic disc. There is a differential ideal $\mathcal{J} \subset \Omega_{\mathbf{R}}^*(\mathbf{C}^n)$ whose integral manifolds are precisely holomorphic curves, defined by

$$\mathcal{J} = \{(\Omega^{2,0}(\mathbf{C}^n) + \Omega^{0,2}(\mathbf{C}^n)) \cap \Omega_{\mathbf{R}}^2(\mathbf{C}^n)\}.$$

In other words, $\mathcal{J}$ is algebraically generated by real 2-forms which, when regarded as complex 2-forms, have no part of type $(1,1)$. It is elementary to see that in degree $k \geq 3$, $\mathcal{J}^k = \Omega_{\mathbf{R}}^k(\mathbf{C}^n)$, and that the integral 2-planes in $T\mathbf{C}^n$ are exactly the complex 1-dimensional subspaces. This implies our claim that integral manifolds of $\mathcal{J}$ are holomorphic curves.

Now, $\mathcal{J}$ has many conservation laws. Namely, for any holomorphic 1-form $\varphi \in \Omega_{hol}^{1,0}(\mathbf{C}^n)$, we find that

$$d\varphi + d\bar\varphi \in \mathcal{J},$$

so that $\varphi + \bar{\varphi}$ is a conservation law for $\mathcal{J}$. These give rise to infinitely many moment conditions

$$\int_{\gamma_{\mathbf{C}}} \varphi = 0$$

which must be satisfied by $\gamma_{\mathbf{C}}$, if it is to be the boundary of a holomorphic disc.

It is a fact which we shall not prove here that *every* conservation law for $\mathcal{J}$ is of this form; trivial conservation laws clearly arise when $\varphi = df$ for some holomorphic function $f \in \mathcal{O}(\mathbf{C}^n)$. Another fact, not to be proved here, is that these moment conditions are sufficient for $\gamma_{\mathbf{C}}$ to bound a (possibly branched) holomorphic disc.

Returning to our discussion of minimal surfaces, suppose that $x : U \to \mathbf{E}^3$ is a minimal immersion of a simply connected surface. Then $*dx$ defines a closed, vector-valued 1-form on U, so there exists a vector-valued function $y : U \to \mathbf{E}^3$ satisfying

$$dy = *dx. \tag{1.21}$$

Note that our ability to integrate the conservation law to obtain a function relies essentially on the fact that we are in dimension $n = 2$.

We can define

$$z = (x + iy) : U \to \mathbf{C}^3,$$

and (1.21) is essentially the Cauchy-Riemann equations, implying that z is a holomorphic curve, with the conformal structure induced from the immersion z. Furthermore, the complex derivative z' is at each point of U a null vector for the complex bilinear inner-product $\sum (dz^i)^2$. This gives the classical Weierstrass representation of a minimal surface in $\mathbf{E}^3$ as locally the real part of a holomorphic null curve in $\mathbf{C}^3$.

We can now incorporate the result of our digression on conservation laws for holomorphic discs. Namely, given a strip $\gamma^{(1)}$, the Euclidean moment condition $\int_{\gamma} *dx = 0$ implies that there exists another real curve y so that $dy = *dx$ (along γ). Then we can use $z = (x + iy) : \gamma \to \mathbf{C}^3$ as initial data for the holomorphic disc problem, and all of the holomorphic moment conditions for that problem come into play. These are the additional hidden constraints needed to fill the real curve γ with a (possibly branched) minimal surface.

Conservation Laws for Constant Mean Curvature

It is also a worthwhile exercise to determine the conservation laws corresponding to Euclidean motions for the constant mean curvature system when the constant H is non-zero. Recall that for that system the Poincaré-Cartan form

$$\Pi_H = -\theta \wedge \left(\sum \pi_i \wedge \omega_{(i)} - H\omega \right)$$

is invariant under the full Euclidean group, but that no particular Lagrangian Λ is so invariant; we will continue to work with the $SO(n + 1, \mathbf{R})$-invariant Lagrangian

$$\Lambda_H = \omega + \tfrac{H}{n+1} x \lrcorner \Omega, \qquad d\Lambda_H = \Pi_H.$$

Fortunately, the equivalence class $[\Lambda] \in H^n(\Omega^*(M)/\mathcal{I})$ *is invariant under the* Euclidean group, because as the reader can verify, the connecting map

$$\delta : H^n(\Omega^*/\mathcal{I}) \to H^{n+1}(\mathcal{I})$$

taking $[\Lambda]$ to Π is an isomorphism for this contact manifold. This means that, as in the case $H = 0$, we will find conservation laws corresponding to the full Euclidean Lie algebra.

Computing the conservation laws corresponding to translations requires the more complicated form of the Noether prescription, because it is the translation vector fields $v \in \mathbf{R}^{n+1}$ which fail to preserve our Λ_H. Instead, we have

$$
\begin{aligned}
\mathcal{L}_v\Lambda_H &= \tfrac{H}{n+1}\mathcal{L}_v(x \lrcorner \Omega) \text{ (because } \mathcal{L}_v\omega = 0), \\
&= \tfrac{H}{n+1}((\mathcal{L}_v x) \lrcorner \Omega + x \lrcorner (\mathcal{L}_v\Omega)) \\
&= \tfrac{H}{n+1}(v \lrcorner \Omega + 0).
\end{aligned}
$$

In the last step, we have used $\mathcal{L}_v x = [v, x] = v$ (by a simple calculation), and $\mathcal{L}_v\Omega = 0$ (because the ambient volume Ω is translation invariant). To apply the Noether prescription, we need an anti-derivative of this last term, which we find by experimenting:

$$
\begin{aligned}
d(x \lrcorner (v \lrcorner \Omega)) &= \mathcal{L}_x(v \lrcorner \Omega) - x \lrcorner d(v \lrcorner \Omega) \\
&= ((\mathcal{L}_x v) \lrcorner \Omega + v \lrcorner (\mathcal{L}_x\Omega)) - x \lrcorner 0 \\
&= -v \lrcorner \Omega + (n + 1)v \lrcorner \Omega,
\end{aligned}
$$

where we have again used $\mathcal{L}_x v = [x, v] = -v$, and $\mathcal{L}_x\Omega = (n + 1)\Omega$. Combining these two calculations, we have

$$\mathcal{L}_v\Lambda_H = \tfrac{H}{n(n+1)}d(x \lrcorner (v \lrcorner \Omega)).$$

The prescription (1.13, 1.14, 1.15) now gives

$$
\begin{aligned}
\varphi_v &= -v \lrcorner \Lambda_H + \tfrac{H}{n(n+1)}x \lrcorner (v \lrcorner \Omega) \\
&= -v \lrcorner \omega + \tfrac{H}{n}x \lrcorner (v \lrcorner \Omega).
\end{aligned}
$$

As in the case of minimal hypersurfaces, we consider the restriction of φ_v to an integral manifold N. From the previous case, we know that $v \lrcorner \omega$ restricts to $*\langle v, dx \rangle$, where $* = *_N$ is the star operator of the metric on N and $\langle \cdot, \cdot \rangle$ denotes the ambient inner-product. To express the restriction of the other term of φ_v, decompose $x = x_t + x_\nu\nu$ into tangential and normal parts along N (so x_t is a vector and x_ν is a scalar), and a calculation gives

$$
\begin{aligned}
\tfrac{H}{n}x \lrcorner (v \lrcorner \Omega)|_N &= -\tfrac{H}{n}(x_\nu v \lrcorner \omega - (v \lrcorner \theta)(x_t \lrcorner \omega)) \\
&= -\tfrac{H}{n}(x_\nu *\langle v, dx \rangle - (v \lrcorner \theta) *\langle \cdot, x_t \rangle);
\end{aligned}
$$

the latter $*$ is being applied to the 1-form on N that is dual via the metric to the tangent vector x_t. Again as in the $H = 0$ case, we can write these $(n - 1)$-forms

φ_v, which depend linearly on $v \in \mathbf{R}^{n+1}$, as an $(\mathbf{R}^{n+1})^*$-valued $(n-1)$-form on N. It is

$$\varphi_{trans} = -(1 + \tfrac{H}{n}x_\nu) *dx + \tfrac{H}{n}\nu *\langle \cdot, x_t \rangle.$$

In the second term, the normal ν provides the "vector-valued" part (it replaced θ, to which it is dual), and $*\langle \cdot, x_t \rangle$ provides the "$(n-1)$-form" part.

Calculating the conservation laws for rotations is a similar process, simplified somewhat by the fact that $\mathcal{L}_v\Lambda_H = 0$; of course, the lifted rotation vector fields v are not so easy to work with as the translations. The resulting $\Lambda^2\mathbf{R}^{n+1}$-valued $(n-1)$-form is

$$\varphi_{rot} = -(1 + \tfrac{H}{n+1}x_\nu) *(x \wedge dx) + \tfrac{H}{n+1}(x \wedge \nu) *\langle \cdot, x_t \rangle.$$

These can be used to produce moment conditions, just as in the $H = 0$ case.

We conclude with one more observation suggesting extensions of the notion of a conservation law. Recall that we showed in (1.7) that a Monge-Ampère system $\mathcal{E}_\Lambda$ might have an infinitesimal symmetry which scales the corresponding Poincaré-Cartan form Π. This is the case for the minimal surface system, which is preserved by the dilation vector field on $\mathbf{E}^{n+1}$

$$x = \sum x^a \tfrac{\partial}{\partial x^a}.$$

This induces a vector field x on the contact manifold of tangent hyperplanes to $\mathbf{E}^{n+1}$ where the functional Λ and Poincaré-Cartan form Π are defined, and there are various ways to calculate that

$$\mathcal{L}_x\Lambda = n\Lambda.$$

If one tries to apply the Noether prescription to x by writing

$$\varphi_{dil} = x \lrcorner \Lambda,$$

the resulting form satisfies

$$\begin{aligned} d\varphi_{dil} &= \mathcal{L}_x\Lambda - x \lrcorner d\Lambda \\ &= n\Lambda - x \lrcorner \Pi. \end{aligned}$$

Restricted to a minimal surface N, we will then have

$$d\varphi_{dil}|_N = n\Lambda.$$

Because the right-hand side is not zero, we do not have a conservation law, but it is still reasonable to look for consequences of integrating on neighborhoods U in N, where we find

$$\int_{\partial U} \varphi = n \int_U \Lambda. \tag{1.22}$$

The right-hand side equals n times the area of U, and the left-hand side can be investigated by choosing an oriented orthonormal frame field $(e_0, \ldots, e_n)$ along $U \cup \partial U$ satisfying the conditions (1.20) as before. We write the coefficients

$$x = \sum x^a \tfrac{\partial}{\partial x^a} = \sum v^a e_a,$$

and then restricted to ∂U, we have

$$\varphi|_{\partial U} = x \,\lrcorner\, \Lambda = (-1)^{n-1} v^n \omega^1 \wedge \cdots \wedge \omega^{n-1}.$$

Up to sign, the form $\omega^1 \wedge \cdots \wedge \omega^{n-1}$ along ∂U is exactly the $(n-1)$-dimensional area form for ∂U.

These interpretations of the two sides of (1.22) can be exploited by taking for U the family of neighborhoods U_r for $r > 0$, defined as the intersection of $N \subset \mathbf{E}^{n+1}$ with an origin-centered ball of radius r. In particular, along ∂U_r we will have $\|x\| = r$, so that $v^n \leq r$ and

$$r \cdot \mathrm{Area}(\partial U_r) \geq n \cdot \mathrm{Vol}(U_r). \tag{1.23}$$

Observe that

$$\mathrm{Area}(\partial U_r) = \frac{d}{dr} \mathrm{Vol}(U_r),$$

and (1.23) is now a differential inequality for $\mathrm{Vol}(U_r)$ which can be solved to give

$$\mathrm{Vol}(U_r) \geq C r^n$$

for some constant C. This is a remarkable result about minimal hypersurfaces, and amply illustrates the power of "almost-conservation laws" like φ_{dil}.

Chapter 2

The Geometry of Poincaré-Cartan Forms

In this chapter, we will study some of the geometry associated to Poincaré-Cartan forms using É. Cartan's method of equivalence. The idea is to identify such a Poincaré-Cartan form with a G-structure—that is, a subbundle of the principal coframe bundle of a manifold—and then attempt to find some canonically determined basis of 1-forms on the total space of that G-structure. The differential structure equations of these 1-forms will then exhibit associated geometric objects and invariants.

The pointwise linear algebra of a Poincaré-Cartan form in the case of $n = 2$ "independent variables" (that is, on a contact manifold of dimension 5) is quite different from that of higher-dimensional cases. Therefore, in the first section we study only the former, which should serve as a good illustration of the method of equivalence for those not familiar with it. Actually, in case $n = 2$ we will study the coarser equivalence of Monge-Ampere systems rather than Poincaré-Cartan forms, and we will do this without restricting to those systems which are locally Euler-Lagrange. An extensive study of the geometry of Monge-Ampere systems in various low dimensions was carried out in [LRC93], with a viewpoint somewhat similar to ours.

In the succeeding sections, we will first identify in case $n \geq 3$ a narrower class of Poincaré-Cartan forms, called *neo-classical*, which are of the same algebraic type as those arising from classical variational problems. We will describe some of the geometry associated with neo-classical Poincaré-Cartan forms, consisting of a field of hypersurfaces in a vector bundle, well-defined up to fiberwise affine motions of the vector bundle. A digression on the local geometry of individual hypersurfaces in affine space follows this. We then turn to the very rich equivalence problem for neo-classical Poincaré-Cartan forms; the differential invariants that this uncovers include those of the various associated affine hypersurfaces. In the last section of this chapter, we use these differential invariants to characterize systems locally contact-equivalent to those for prescribed mean curvature hypersurfaces in Riemannian manifolds.

In the next chapter, we will specialize to the study of those neo-classical Poincaré-Cartan forms whose primary differential invariants all vanish. These correspond to interesting variational problems arising in conformal geometry.

We begin with a few elementary notions used in the method of equivalence. On a manifold M of dimension n, a *coframe* at a point $x \in M$ is a linear isomorphism

$$u_x : T_x M \xrightarrow{\sim} \mathbf{R}^n.$$

This is equivalent to a choice of basis for the cotangent space $T_x^* M$, and we will not maintain any distinction between these two notions. The set of all coframes for M has the structure of a principal $GL(n, \mathbf{R})$-bundle $\pi : \mathcal{F}(M) \to M$, with right-action

$$u_x \cdot g \stackrel{def}{=} g^{-1} u_x, \qquad g \in GL(n, \mathbf{R}),$$

where the right-hand side denotes composition of u_x with multiplication by g^{-1}. A local section of $\pi : \mathcal{F}(M) \to M$ is called a *coframing*, or *coframe field*. On the total space $\mathcal{F}(M)$, there is an $\mathbf{R}^n$-valued *tautological 1-form* ω, given at $u \in \mathcal{F}(M)$ by

$$\omega_u(v) = u(\pi_* v) \in \mathbf{R}^n, \qquad v \in T_u \mathcal{F}(M). \tag{2.1}$$

The n components ω^i of this $\mathbf{R}^n$-valued 1-form give a global basis for the semibasic 1-forms of $\mathcal{F}(M) \to M$.

In terms of coordinates $x = (x^1, \ldots, x^n)$ on M, there is a trivialization $M \times GL(n, \mathbf{R}) \cong \mathcal{F}(M)$ given by

$$(x, g) \leftrightarrow (x, g^{-1} dx),$$

where on the right-hand side, dx is a column of 1-forms regarded as a coframe at x, and $g^{-1} dx$ is the composition of that coframe with multiplication by $g^{-1} \in GL(n, \mathbf{R})$. In this trivialization, we can express the tautological 1-form as

$$\omega = g^{-1} dx,$$

where again the right-hand side represents the product of a $GL(n, \mathbf{R})$-valued fiber coordinate and an $\mathbf{R}^n$-valued semibasic 1-form.

The geometric setting of the equivalence method is the following.

Definition 2.1 *Let $G \subset GL(n, \mathbf{R})$ be a subgroup. A G-structure on the n-manifold M is a principal subbundle of the coframe bundle $\mathcal{F}(M) \to M$, having structure group G.*

We will associate to a hyperbolic Monge-Ampère system (to be defined, in case $n = 2$), or to a neo-classical Poincaré-Cartan form (in case $n \geq 3$), a succession of G-structures on the contact manifold M, which carry increasingly detailed information about the geometry of the system or form, respectively.

2.1 The Equivalence Problem for $n = 2$

In this section, we will study the equivalence problem for certain Monge-Ampere systems on contact manifolds of dimension 5. We will give criteria in terms of the differential invariants thus obtained for a given system to be locally equivalent to the system associated to the linear homogeneous wave equation. We will also give the weaker criteria for a given system to be locally equivalent to an Euler-Lagrange system, as in the previously discussed inverse problem. Unless otherwise noted, we use the index ranges $0 \leq a, b, c \leq 4$, $1 \leq i, j, k \leq 4$.

We assume given a 5-dimensional contact manifold (M, I) and a Monge-Ampere system $\mathcal{E}$, locally algebraically generated as

$$\mathcal{E} = \{\theta, d\theta, \Psi\},$$

where $0 \neq \theta \in \Gamma(I)$ is a contact form, and $\Psi \in \Omega^2(M)$ is some 2-form. As noted previously, $\mathcal{E}$ determines I and $\mathcal{I}$. We assume that $\Psi_x \notin \mathcal{I}_x$ for all $x \in M$. Recall from the discussion in §1.2.3 that given $\mathcal{E}$, the generator Ψ may be uniquely chosen modulo $\{I\}$ (and modulo multiplication by functions) by the condition of primitivity; that is, we may assume

$$d\theta \wedge \Psi \equiv 0 \pmod{\{I\}}.$$

The assumption $\Psi_x \notin \mathcal{I}_x$ means that this primitive form is non-zero everywhere. We do *not* necessarily assume that $\mathcal{E}$ is Euler-Lagrange.

On the contact manifold M, one can locally find a coframing $\eta = (\eta^a)$ such that

$$\begin{cases} \eta^0 \in \Gamma(I), \\ d\eta^0 \equiv \eta^1 \wedge \eta^2 + \eta^3 \wedge \eta^4 \pmod{\{I\}}. \end{cases} \tag{2.2}$$

Then we can write $\Psi \equiv \frac{1}{2} b_{ij} \eta^i \wedge \eta^j \pmod{\{I\}}$, where the functions b_{ij} depend on the choice of coframing and on the choice of Ψ. The assumption that Ψ is primitive means that in terms of a coframing satisfying (2.2),

$$b_{12} + b_{34} = 0.$$

We now ask what further conditions may be imposed on the coframing $\eta = (\eta^a)$ while preserving (2.2).

To investigate this, we first consider changes of coframe that fix η^0; we will later take into account non-trivial rescalings of η^0. In this case, an element of $GL(5, \mathbf{R})$ preserves the condition (2.2) if and only if it acts as a fiberwise *sympletic* transformation, modulo the contact line bundle I. Working modulo I, we can split

$$\bigwedge^2(T^*M/I) \cong (\mathbf{R} \cdot d\eta^0) \oplus P^2(T^*M/I),$$

where $P^2(T^*M/I)$ is the 5-dimensional space of 2-forms that are primitive with respect to the symplectic structure on $I^\perp$ induced by $d\eta^0$. The key observation is that the action of the symplectic group $Sp(2, \mathbf{R})$ on $P^2(\mathbf{R}^4)$ is equivalent to the standard action of the group $SO(3,2)$ on $\mathbf{R}^5$. This is because the symmetric bilinear form $\langle \cdot, \cdot \rangle$ on $P^2(T^*M/I)$ defined by

$$\psi_1 \wedge \psi_2 = \langle \psi_1, \psi_2 \rangle (d\eta^0)^2$$

has signature $(3,2)$ and symmetry group $Sp(2,\mathbf{R})$. Therefore, the orbit decomposition of the space of primitive forms Ψ modulo $\{I\}$ under admissible changes of coframe will be a refinement of the standard orbit decomposition under $SO(3,2)$.

To incorporate rescaling of η^0 into our admissible changes of coframe, note that a rescaling of η^0 requires via (2.2) the same rescaling of the symplectic form $\eta^1 \wedge \eta^2 + \eta^3 \wedge \eta^4$, so we should actually allow changes by elements of $GL(5,\mathbf{R})$ inducing the standard action of $CSp(2,\mathbf{R})$; this is the group that preserves the standard symplectic form up to scale. This in turn corresponds to the split-signature conformal group $CO(3,2)$, which acts on $\mathbf{R}^5$ with three non-zero orbits: a negative space, a null space, and a positive space.

The three orbits of this representation correspond to three types of Monge-Ampere systems:

- If $\Psi \wedge \Psi$ is a negative multiple of $d\eta^0 \wedge d\eta^0$, then the local coframing η may be chosen so that in addition to (2.2),

$$\Psi \equiv \eta^1 \wedge \eta^2 - \eta^3 \wedge \eta^4 \pmod{\{I\}};$$

for a classical variational problem, this occurs when the Euler-Lagrange PDE is hyperbolic.

- If $\Psi \wedge \Psi = 0$, then η may be chosen so that

$$\Psi \equiv \eta^1 \wedge \eta^3 \pmod{\{I\}};$$

for a classical variational problem, this occurs when the Euler-Lagrange PDE is parabolic.

- If $\Psi \wedge \Psi$ is a positive multiple of $d\eta^0 \wedge d\eta^0$, then η may be chosen so that

$$\Psi \equiv \eta^1 \wedge \eta^4 - \eta^3 \wedge \eta^2 \pmod{\{I\}};$$

for a classical variational problem, this occurs when the Euler-Lagrange PDE is elliptic.

The equivalence problem for elliptic Monge-Ampere systems in case $n = 2$ develops in analogy with that for hyperbolic systems; we will present the hyperbolic case. The conclusion will be:

Associated to a hyperbolic Monge-Ampere system $(M^5, \mathcal{E})$ is a canonical subbundle $B_1 \to M$ of the coframe bundle of M carrying a pair of 2×2-matrix-valued functions S_1 and S_2, involving up to second derivatives of the given system. $(M, \mathcal{E})$ is locally of Euler-Lagrange type if and only if S_2 vanishes identically, while it is equivalent to the system associated to the homogeneous wave equation $z_{xy} = 0$ if and only if S_1 and S_2 both vanish identically.

An example of a hyperbolic Monge-Ampere system, to be studied in more detail in Chapter 4, is the linear Weingarten system for surfaces in $\mathbf{E}^3$ with Gauss curvature $K = -1$.

To begin, assume that $(M^5, \mathcal{E})$ is a hyperbolic Monge-Ampere system. A coframing $\eta = (\eta^a)$ of M is said to be 0-*adapted* to $\mathcal{E}$ if

$$\mathcal{E} = \{\eta^0,\ \eta^1 \wedge \eta^2 + \eta^3 \wedge \eta^4,\ \eta^1 \wedge \eta^2 - \eta^3 \wedge \eta^4\} \tag{2.3}$$

and also

$$d\eta^0 \equiv \eta^1 \wedge \eta^2 + \eta^3 \wedge \eta^4 \pmod{\{I\}}. \tag{2.4}$$

According to the following proposition, a hyperbolic Monge-Ampere system is equivalent to a certain type of G-structure, and it is the latter to which the equivalence method directly applies.

Proposition 2.1 *The 0-adapted coframings for a hyperbolic Monge-Ampere system* $(M^5, \mathcal{E})$ *are the sections of a* G_0-*structure on* M, *where* $G_0 \subset GL(5, \mathbf{R})$ *is the (disconnected) subgroup generated by all matrices of the form (displayed in blocks of size* $1, 2, 2$)

$$g_0 = \begin{pmatrix} a & 0 & 0 \\ C & A & 0 \\ D & 0 & B \end{pmatrix}, \tag{2.5}$$

with $a = det(A) = det(B) \neq 0$, *along with the matrix*

$$J = \begin{pmatrix} 1 & 0 & 0 \\ 0 & 0 & I_2 \\ 0 & I_2 & 0 \end{pmatrix}. \tag{2.6}$$

Proof. The content of this proposition is that any two 0-adapted coframes differ by multiplication by an element of G_0. To see why this is so, note that the 2-forms

$$\eta^1 \wedge \eta^2 + \eta^3 \wedge \eta^4 \quad \text{and} \quad \eta^1 \wedge \eta^2 - \eta^3 \wedge \eta^4$$

have, up to scaling, exactly 2 decomposable linear combinations, $\eta^1 \wedge \eta^2$ and $\eta^3 \wedge \eta^4$. These must be either preserved or exchanged by any change of coframe preserving their span modulo $\{I\}$, and this accounts for both the block form (2.5) and the matrix J. The condition on determinants then corresponds to (2.4). □

Although not every G_0-structure on a 5-manifold M is induced by a hyperbolic Monge-Ampere system $\mathcal{E}$, it is easy to see that those that do determine $\mathcal{E}$ uniquely. We therefore make a digression to describe the first steps of the equivalence method, by which one investigates the local geometry of a general G-structure. This will be followed by application to the case at hand of a G_0-structure, then a digression on the next general steps, and application to the case at hand, and so on. One major step, that of *prolongation*, will not appear in this chapter but will be discussed in the study of conformal geometry in Chapter 3.

Fix a subgroup $G \subset GL(n, \mathbf{R})$. Two G-structures $B_i \to M_i$, $i = 1, 2$, are *equivalent* if there is a diffeomorphism $M_1 \to M_2$ such that under the induced isomorphism of principal coframe bundles $\mathcal{F}(M_1) \to \mathcal{F}(M_2)$, the subbundle $B_1 \subset \mathcal{F}(M_1)$ is mapped to $B_2 \subset \mathcal{F}(M_2)$. One is typically interested only in those properties of a G-structure which are preserved under this notion of equivalence. For instance, if one has a pair of 5-manifolds with hyperbolic Monge-Ampere systems, then a diffeomorphism of the 5-manifolds carries one of these systems to the other if and only if it induces an equivalence of the associated G_0-structures.

It is easy to see that a diffeomorphism $F : B_1 \to B_2$ between the total spaces of two G-structures $B_i \to M_i$ is an equivalence in the above sense if and only if $F^*(\omega_2) = \omega_1$, where ω_i is the restriction of the tautological $\mathbf{R}^n$-valued form (2.1) on $\mathcal{F}(M_i) \supseteq B_i$. The first step in investigating the geometry of a G-structure $B \to M$ is therefore to understand the local behavior of this tautological form. To do this, we seek an expression for its exterior derivative, and to understand what such an expression should look like, we proceed as follows.

Consider a local trivialization $B \cong M \times G$, induced by a choice of section η of $B \to M$ whose image is identified with $M \times \{e\} \subset M \times G$. The section η is in particular an $\mathbf{R}^n$-valued 1-form on M, and the tautological 1-form is

$$\omega = g^{-1}\eta \in \Omega^1(B) \otimes \mathbf{R}^n.$$

The exterior derivative of this equation is

$$d\omega = -g^{-1}dg \wedge \omega + g^{-1}d\eta. \tag{2.7}$$

Note that the last term in this equation is semibasic for $B \to M$, and that the matrix 1-form $g^{-1}dg$ takes values in the Lie algebra $\mathfrak{g}$ of G. Of course, these pieces $g^{-1}d\eta$ and $g^{-1}dg$ each depend on the choice of trivialization. To better understand the pointwise linear algebra of (2.7), we introduce the following notion.

Definition 2.2 *A* pseudo-connection *in the G-structure $B \to M$ is a $\mathfrak{g}$-valued 1-form on B whose restriction to the fiber tangent spaces $V_b \subset T_bB$ equals the identification $V_b \cong \mathfrak{g}$ induced by the right G-action on B.*

This differs from the definition of a *connection* in the principal bundle $B \to M$ by omission of an equivariance requirement. In terms of our trivialization above, a pseudo-connection on $M \times G$ is any $\mathfrak{g}$-valued 1-form of the form

$$g^{-1}dg + (\text{semibasic } \mathfrak{g}\text{-valued 1-form});$$

in particular, every G-structure carries a pseudo-connection. A consequence of (2.7) is that any pseudo-connection $\varphi \in \Omega^1(B) \otimes \mathfrak{g}$ satisfies a structure equation that is fundamental for the equivalence method:

$$\boxed{d\omega = -\varphi \wedge \omega + \tau,} \tag{2.8}$$

where $\tau = (\frac{1}{2}T^i_{jk}\omega^j \wedge \omega^k)$ is a semibasic $\mathbf{R}^n$-valued 2-form on B, called the *torsion* of the pseudo-connection φ. It is natural to consider exactly how a different choice of pseudo-connection—remember that any two differ by an arbitrary semibasic $\mathfrak{g}$-valued 1-form—yields a different torsion form. We will pursue this after considering the situation for our hyperbolic Monge-Ampere systems.

Let $B_0 \subset \mathcal{F}(M)$ be the G_0-bundle of 0-adapted coframes for a hyperbolic Monge-Ampere system $\mathcal{E}$. A local section η corresponds to an $\mathbf{R}^5$-valued 1-form (η^a) satisfying (2.3, 2.4). In terms of the trivialization $B_0 \cong M \times G_0$ induced by η, the tautological $\mathbf{R}^5$-valued 1-form is $\omega = g_0^{-1}\eta$. Locally (over neighborhoods in M), there is a structure equation (2.8), in which

$$\omega = \begin{pmatrix} \omega^0 \\ \omega^1 \\ \omega^2 \\ \omega^3 \\ \omega^4 \end{pmatrix} \quad \text{and} \quad \varphi = \begin{pmatrix} \varphi^0_0 & 0 & 0 & 0 & 0 \\ \varphi^1_0 & \varphi^1_1 & \varphi^1_2 & 0 & 0 \\ \varphi^2_0 & \varphi^2_1 & \varphi^2_2 & 0 & 0 \\ \varphi^3_0 & 0 & 0 & \varphi^3_3 & \varphi^3_4 \\ \varphi^4_0 & 0 & 0 & \varphi^4_3 & \varphi^4_4 \end{pmatrix}$$

are the tautological $\mathbf{R}^5$-valued 1-form and the pseudo-connection form, respectively; note that the condition for φ to be $\mathfrak{g}_0$-valued includes the condition

$$\varphi^0_0 = \varphi^1_1 + \varphi^2_2 = \varphi^3_3 + \varphi^4_4.$$

The torsion τ of φ is an $\mathbf{R}^5$-valued 2-form, semibasic for $B_0 \to M$ and depending on a choice of pseudo-connection.

Returning to the general situation of a G-structure $B \to M$, our goal is to understand how different choices of pseudo-connection in (2.8) yield different torsion forms. We will use this to restrict attention to those pseudo-connections whose torsion is in some normal form.

The linear-algebraic machinery for this is as follows. Associated to the linear Lie algebra $\mathfrak{g} \subset \mathfrak{gl}(n, \mathbf{R})$ is a map of G-modules

$$\delta : \mathfrak{g} \otimes (\mathbf{R}^n)^* \to \mathbf{R}^n \otimes \textstyle\bigwedge^2(\mathbf{R}^n)^*,$$

defined as the restriction to

$$\mathfrak{g} \otimes (\mathbf{R}^n)^* \subset (\mathbf{R}^n \otimes (\mathbf{R}^n)^*) \otimes (\mathbf{R}^n)^* \tag{2.9}$$

of the surjective skew-symmetrization map

$$\mathbf{R}^n \otimes (\mathbf{R}^n)^* \otimes (\mathbf{R}^n)^* \to \mathbf{R}^n \otimes \textstyle\bigwedge^2(\mathbf{R}^n)^*.$$

The cokernel of δ

$$H^{0,1}(\mathfrak{g}) \stackrel{\text{def}}{=} (\mathbf{R}^n \otimes \textstyle\bigwedge^2(\mathbf{R}^n)^*)/\delta(\mathfrak{g} \otimes (\mathbf{R}^n)^*) \tag{2.10}$$

is one of the *Spencer cohomology groups* of $\mathfrak{g} \subset \mathfrak{gl}(n, \mathbf{R})$. Note that to each $b \in B$ is associated an isomorphism $T_{\pi(b)}M \xrightarrow{\sim} \mathbf{R}^n$, and consequently an identification of semibasic 1-forms at $b \in B$ with $(\mathbf{R}^n)^*$. Now, given a pseudo-connection in the G-structure, the semibasic $\mathbf{R}^n$-valued torsion 2-form $(\frac{1}{2}T^i_{jk}\omega^j \wedge \omega^k)$ at $b \in B$ can be identified with an element $\tau_b \in \mathbf{R}^n \otimes \Lambda^2(\mathbf{R}^n)^*$. Similarly, a permissible change at $b \in B$ of the pseudo-connection—that is, a semibasic $\mathfrak{g}$-valued 1-form—can be identified with an element $\varphi'_b \in \mathfrak{g} \otimes (\mathbf{R}^n)^*$. Under these two identifications, the map δ associates to a change φ'_b the corresponding change in the torsion $\varphi'_b \wedge \omega_b$, where in this expression we have contracted the middle factor of $(\mathbf{R}^n)^*$ in φ'_b (see (2.9)) with the values of the $\mathbf{R}^n$-valued 1-form ω_b. Therefore, different choices of pseudo-connection yield torsion maps differing by elements of $Im(\delta)$, so what is determined by the G-structure alone, independent of a choice of pseudo-connection, is a map $\bar{\tau}: B \to H^{0,1}(\mathfrak{g})$, called the *intrinsic torsion* of $B \to M$.

This suggests a major step in the equivalence method, called *absorption of torsion*, which one implements by choosing a (vector space) splitting of the projection

$$\mathbf{R}^n \otimes \bigwedge\nolimits^2(\mathbf{R}^n)^* \to H^{0,1}(\mathfrak{g}) \to 0. \tag{2.11}$$

As there may be no G-equivariant splitting, one is merely choosing some vector subspace $T \subset \mathbf{R}^n \otimes \bigwedge^2(\mathbf{R}^n)^*$ which complements the kernel $\delta(\mathfrak{g} \otimes (\mathbf{R}^n)^*)$. Fixing a choice of T, it holds by construction that any G-structure $B \to M$ locally has pseudo-connections whose torsion at each $b \in B$ corresponds to a tensor lying in T.

We will see from our example of hyperbolic Monge-Ampere systems that this is not as complicated as it may seem. Denote the semibasic 2-form components of the $\mathbf{R}^5$-valued torsion by

$$\tau = \begin{pmatrix} \tau^0 \\ \tau^1 \\ \tau^2 \\ \tau^3 \\ \tau^4 \end{pmatrix}.$$

We know from the condition (2.4) in the definition of 0-adapted that

$$\tau^0 \stackrel{def}{=} d\omega^0 + \varphi^0_0 \wedge \omega^0 = \omega^1 \wedge \omega^2 + \omega^3 \wedge \omega^4 + \sigma \wedge \omega^0$$

for some semibasic 1-form σ. We may now replace φ^0_0 by $\varphi^0_0 - \sigma$ in our pseudo-connection, eliminating the term $\sigma \wedge \omega^0$ from the torsion. We then rename this altered pseudo-connection entry again as φ^0_0; to keep the pseudo-connection $\mathfrak{g}_0$-valued, we have to make a similar change in $\varphi^1_1 + \varphi^2_2$ and $\varphi^3_3 + \varphi^4_4$. What we have just shown is that given an arbitrary pseudo-connection in a G_0-structure $B_0 \to M$, there is another pseudo-connection whose torsion satisfies (using obvious coordinates on $\mathbf{R}^n \otimes \bigwedge^2(\mathbf{R}^n)^*$) $T^0_{0a} = T^0_{a0} = 0$. By choosing this latter pseudo-connection, we are absorbing the corresponding torsion components into φ. Furthermore, the fact that our G_0-structure is not arbitrary, but comes from a hyperbolic Monge-Ampère system, gave us the additional information that $T_{12}^0 = T_{34}^0 = 1$, and all other independent $T_{ij}^0 = 0$. Note incidentally that our decision to use a pseudo-connection giving $\sigma = 0$ determines φ_0^0 uniquely, up to addition of multiples of ω^0; this uniqueness applies also to $\varphi_1^1 + \varphi_2^2$ and $\varphi_3^3 + \varphi_4^4$. The effort to uniquely determine pseudo-connection forms should guide the choices one makes in the equivalence method.

Other torsion terms may be absorbed using similar methods. Using the index range $1 \leq i, j, k \leq 4$, we write

$$\tau^i = T_{j0}^i \omega^j \wedge \omega^0 + \tfrac{1}{2} T_{jk}^i \omega^j \wedge \omega^k$$

for functions T_{j0}^i and $T_{jk}^i = -T_{kj}^i$. First, by altering the nilpotent part φ_0^i, we can arrange that all $T_{j0}^i = 0$. Second, by altering the off-diagonal terms φ_2^1, φ_1^2, φ_3^4, φ_4^3, we can arrange that

$$T_{2j}^1 = T_{1j}^2 = T_{4j}^3 = T_{3j}^4 = 0.$$

Third, by altering the traceless diagonal parts $\varphi_1^1 - \varphi_2^2$ and $\varphi_3^3 - \varphi_4^4$, we can arrange that

$$T_{13}^1 = T_{23}^2, \ T_{14}^1 = T_{24}^2, \ T_{13}^3 = T_{14}^4, \ T_{23}^3 = T_{24}^4.$$

We summarize this by renaming

$$
\begin{aligned}
\tau^1 &= (V_3 \omega^3 + V_4 \omega^4) \wedge \omega^1 + U^1 \omega^3 \wedge \omega^4, \\
\tau^2 &= (V_3 \omega^3 + V_4 \omega^4) \wedge \omega^2 + U^2 \omega^3 \wedge \omega^4, \\
\tau^3 &= (V_1 \omega^1 + V_2 \omega^2) \wedge \omega^3 + U^3 \omega^1 \wedge \omega^2, \\
\tau^4 &= (V_1 \omega^1 + V_2 \omega^2) \wedge \omega^4 + U^4 \omega^1 \wedge \omega^2,
\end{aligned}
$$

for 8 torsion functions U_i, V_i on B_0. The collection of torsion tensors (T_{bc}^a) taking this form, and satisfying $T_{0a}^0 = T_{a0}^0 = 0$, constitutes the splitting of (2.11) given in the general discussion, to which we will return shortly.

At this point, we can uncover more consequences of the fact that we are dealing not with an arbitrary G_0-structure on a 5-manifold, but a special one induced by a hyperbolic Monge-Ampère system. We already found as one consequence the fact that

$$\tau^0 \equiv \omega^1 \wedge \omega^2 + \omega^3 \wedge \omega^4 \pmod{\{\omega^0\}},$$

which has nothing to do with our choices in absorbing torsion; absorbing torsion allowed us to render this congruence into an equality. Similarly, we now obtain pointwise relations among other torsion coefficients by computing, modulo $\{I\}$

(which in this case means ignoring all ω^0 terms after differentiating),

$$
\begin{aligned}
0 \equiv\ & d(d\omega^0) \\
\equiv\ & \varphi_0^0 \wedge (\omega^1 \wedge \omega^2 + \omega^3 \wedge \omega^4) \\
& + ((-\varphi_1^1 + V_3\omega^3 + V_4\omega^4) \wedge \omega^1 + U^1\omega^3 \wedge \omega^4) \wedge \omega^2 \\
& - \omega^1 \wedge ((-\varphi_2^2 + V_3\omega^3 + V_4\omega^4) \wedge \omega^2 + U^2\omega^3 \wedge \omega^4) \\
& + ((-\varphi_3^3 + V_1\omega^1 + V_2\omega^2) \wedge \omega^3 + U^3\omega^1 \wedge \omega^2) \wedge \omega^4 \\
& - \omega^3 \wedge ((-\varphi_4^4 + V_1\omega^1 + V_2\omega^2) \wedge \omega^4 + U^4\omega^1 \wedge \omega^2) \\
\equiv\ & (U^1 + 2V_2)\omega^2 \wedge \omega^3 \wedge \omega^4 - (U^2 - 2V_1)\omega^1 \wedge \omega^3 \wedge \omega^4 \\
& + (U^3 + 2V_4)\omega^1 \wedge \omega^2 \wedge \omega^4 - (U^4 - 2V_3)\omega^1 \wedge \omega^2 \wedge \omega^3,
\end{aligned}
$$

so that

$$
U^1 = -2V_2,\ U^2 = 2V_1,\ U^3 = -2V_4,\ U^4 = 2V_3.
$$

These are pointwise linear-algebraic relations among our 8 torsion functions.

In the general study of G-structures $B \to M$, we now have to consider the group action in more detail. Specifically, $H^{0,1}(\mathfrak{g})$ is the cokernel of a map of G-modules, so it inherits a G-action as well, and it is easy to see that the intrinsic torsion $\bar{\tau} : B \to H^{0,1}(\mathfrak{g})$ is equivariant for this action. Therefore, there is an induced map

$$
[\bar{\tau}] : M \to H^{0,1}(\mathfrak{g})/G,
$$

which is an invariant of the equivalence class of the G-structure $B \to M$; that is, under a diffeomorphism $M_1 \to M_2$ inducing an equivalence of G-structures, $[\tau_2]$ must pull back to $[\tau_1]$. Now, $H^{0,1}(\mathfrak{g})/G$ typically has a complicated topology, and is rarely a manifold. However, in many cases of interest one can find a *slice* $W \subset H^{0,1}(\mathfrak{g})$, a submanifold whose points all have the same stabilizer $G_1 \subset G$, and which is a cross-section of the orbits which W itself intersects. If the intrinsic torsion $\bar{\tau} : B \to H^{0,1}(\mathfrak{g})$ of a G-structure takes values in a union of orbits represented by such a slice, then the set

$$
B_1 \overset{\text{def}}{=} \bar{\tau}^{-1}(W)
$$

is a smooth principal subbundle of $B \to M$ having structure group $G_1 \subset G$.

The process of reducing to a subbundle defined as the locus where intrinsic torsion lies in a slice is called *normalizing* the torsion. If G_1 is a proper subgroup of G, then we can essentially start the process over, starting with an arbitrary pseudo-connection, absorbing torsion, and so on. Typically, one inherits from $B \to M$ some information about the torsion of the subbundle $B_1 \to M$, because the original structure equations restrict to the submanifold $B_1 \subset B$. We will see an example of this below.

In practice, one typically studies the G-action on $H^{0,1}(\mathfrak{g})$ by transporting it to the representing vector space $T \subset \mathbf{R}^n \otimes \bigwedge^2(\mathbf{R}^n)^*$. If T is not an invariant subspace of $\mathbf{R}^n \otimes \bigwedge^2(\mathbf{R}^n)^*$, then typically G will act by *affine-linear* motions on

T. This is the case in the next step of our equivalence problem for hyperbolic Monge-Ampere systems.

We have represented the intrinsic torsion of a G_0-structure $B_0 \to M$ corresponding to a hyperbolic Monge-Ampere system by 4 independent functions on B; that is, our torsion takes values in a 4-dimensional subspace of the lift T of $H^{0,1}(\mathfrak{g}_0)$. The next step is to determine how the independent torsion functions vary along the fibers of $B_0 \to M$. This will be expressed infinitesimally, in an equation for the exterior derivative of the torsion functions, modulo the space of forms that are semibasic for $B_0 \to M$; the expressions will be in terms of the pseudo-connection forms which parallelize the fibers. They are obtained as follows.

We first consider the equations for $d\omega^1$, $d\omega^2$. Taking the exterior derivative of each, modulo the algebraic ideal $\{\omega^0, \omega^1, \omega^2\}$, yields equivalences of 3-forms that *do not involve derivatives of any pseudo-connection forms*, but do involve dU^1, dU^2. From each of these can be factored the 2-form $\omega^3 \wedge \omega^4$, yielding a pair of equivalences modulo $\{\omega^0, \ldots, \omega^4\}$, expressible in matrix form as

$$0 \equiv d\begin{pmatrix} U^1 \\ U^2 \end{pmatrix} + \begin{pmatrix} \varphi_0^1 \\ \varphi_0^2 \end{pmatrix} + \begin{pmatrix} \varphi_1^1 & \varphi_2^1 \\ \varphi_1^2 & \varphi_2^2 \end{pmatrix} \cdot \begin{pmatrix} U^1 \\ U^2 \end{pmatrix} - \varphi_0^0 \cdot \begin{pmatrix} U^1 \\ U^2 \end{pmatrix}.$$

A similar procedure applied to the equations for $d\omega^3$, $d\omega^4$ yields the pair

$$0 \equiv d\begin{pmatrix} U^3 \\ U^4 \end{pmatrix} + \begin{pmatrix} \varphi_0^3 \\ \varphi_0^4 \end{pmatrix} + \begin{pmatrix} \varphi_3^3 & \varphi_4^3 \\ \varphi_3^4 & \varphi_4^4 \end{pmatrix} \cdot \begin{pmatrix} U^3 \\ U^4 \end{pmatrix} - \varphi_0^0 \cdot \begin{pmatrix} U^3 \\ U^4 \end{pmatrix}.$$

These describe the derivatives of the functions U^i along the fibers of $B_0 \to M$. They are to be interpreted as giving

$$\left.\tfrac{d}{dt}\right|_{t=0} U_i(u \cdot g_t),$$

where g_t is a path in G_0 passing through the identity matrix at $t = 0$. Exponentiated, we see that the vector-valued functions (U_1, U_2) and (U_3, U_4) on B_0 each transform by an *affine-linear action* of G_0 along the fibers; that is, they vary by a linear representation composed with a translation.[1] It is the "nilpotent" part of the group, with components g_0^i, which gives rise to the translation. Specifically, we have for g_0 as in (2.5)

$$\begin{pmatrix} U^1(u \cdot g_0) \\ U^2(u \cdot g_0) \end{pmatrix} = aA^{-1}\begin{pmatrix} U^1(u) \\ U^2(u) \end{pmatrix} - A^{-1}C, \tag{2.12}$$

$$\begin{pmatrix} U^3(u \cdot g_0) \\ U^4(u \cdot g_0) \end{pmatrix} = aB^{-1}\begin{pmatrix} U^3(u) \\ U^4(u) \end{pmatrix} - B^{-1}D. \tag{2.13}$$

---

[1] Strictly speaking, we have only shown that the torsion function (U^i) varies by an affine-linear action under the *identity component* of G_0. What will be important, however, is that if $u \in B_0$ satisfies $U^i(u) = 0$, then $U^i(u \cdot J) = 0$ as well, and likewise for some matrix in each component where $a < 0$. These claims can be verified directly.

Now define a 1-*adapted coframe* to be a 0-adapted coframe $u \in B_0$ satisfying $U^i(u) = 0$ for $1 \le i \le 4$. It then follows from the above reasoning that the subset $B_1 \subset B_0$ of 1-adapted coframes is a G_1-subbudle of B_0, where the subgroup $G_1 \subset G_0$ is generated by the matrix J of (2.6), and by matrices of the form (again, in blocks of size $1, 2, 2$)

$$g_1 = \begin{pmatrix} a & 0 & 0 \\ 0 & A & 0 \\ 0 & 0 & B \end{pmatrix}, \tag{2.14}$$

with $a = \det(A) = \det(B) \ne 0$. The structure equation (2.8) on B_0 still holds when restricted to B_1, with $\tau^i|_{B_1} = 0$; but the pseudo-connection forms $\varphi_0^i|_{B_1}$ are semibasic for $B_1 \to M$, and their contribution should be regarded as torsion. With everything now restricted to B_1, we write

$$\varphi_0^i = P_0^i \omega^0 + P_j^i \omega^j$$

and then have

$$d\omega = -\varphi \wedge \omega + \tau$$

with

$$\varphi = \begin{pmatrix} \varphi_0^0 & 0 & 0 & 0 & 0 \\ 0 & \varphi_1^1 & \varphi_2^1 & 0 & 0 \\ 0 & \varphi_1^2 & \varphi_2^2 & 0 & 0 \\ 0 & 0 & 0 & \varphi_3^3 & \varphi_4^3 \\ 0 & 0 & 0 & \varphi_3^4 & \varphi_4^4 \end{pmatrix} \quad \text{and} \quad \tau = \begin{pmatrix} \omega^1 \wedge \omega^2 + \omega^3 \wedge \omega^4 \\ -P_j^1 \omega^j \wedge \omega^0 \\ -P_j^2 \omega^j \wedge \omega^0 \\ -P_j^3 \omega^j \wedge \omega^0 \\ -P_j^4 \omega^j \wedge \omega^0 \end{pmatrix}.$$

As before, we can absorb some of this torsion into the pseudo-connection form, respecting the constraint $\varphi_0^0 = \varphi_1^1 + \varphi_2^2 = \varphi_3^3 + \varphi_4^4$, until the torsion is of the form

$$d\omega + \varphi \wedge \omega = \begin{pmatrix} \omega^1 \wedge \omega^2 + \omega^3 \wedge \omega^4 \\ -(P\omega^1 + P_3^1 \omega^3 + P_4^1 \omega^4) \wedge \omega^0 \\ -(P\omega^2 + P_3^2 \omega^3 + P_4^2 \omega^4) \wedge \omega^0 \\ -(Q\omega^3 + P_1^3 \omega^1 + P_2^3 \omega^2) \wedge \omega^0 \\ -(Q\omega^4 + P_1^4 \omega^1 + P_2^4 \omega^2) \wedge \omega^0 \end{pmatrix}. \tag{2.15}$$

We can go further: recall that φ_0^0 was uniquely determined up to addition of a multiple of ω^0. We now exploit this, and take the unique choice of $\varphi_0^0 = \varphi_1^1 + \varphi_2^2 = \varphi_3^3 + \varphi_4^4$ that yields a torsion vector of the form (2.15), with

$$P + Q = 0.$$

Now that φ_0^0 is uniquely determined, it is reasonable to try to get information about its exterior derivative. To do this, we differentiate the equation

$$d\omega^0 = -\varphi_0^0 \wedge \omega^0 + \omega^1 \wedge \omega^2 + \omega^3 \wedge \omega^4,$$

which simplifies to

$$
\begin{aligned}
0 = \;& (-d\varphi_0^0 + 2P\omega^1 \wedge \omega^2 + 2Q\omega^3 \wedge \omega^4 \\
& + (P_3^2 - P_1^4)\omega^1 \wedge \omega^3 - (P_3^1 + P_2^4)\omega^2 \wedge \omega^3 \\
& + (P_4^2 + P_1^3)\omega^1 \wedge \omega^4 - (P_4^1 - P_2^3)\omega^2 \wedge \omega^4) \wedge \omega^0.
\end{aligned}
$$

This tells us the derivative of φ_0^0 modulo the algebraic ideal $\{\omega^0\} = \{I\}$, which we now use in a somewhat unintuitive way.

We easily compute that

$$
d(\omega^0 \wedge \omega^1 \wedge \omega^2) = -2\varphi_0^0 \wedge \omega^0 \wedge \omega^1 \wedge \omega^2 + \omega^1 \wedge \omega^2 \wedge \omega^3 \wedge \omega^4.
$$

With knowledge of $d\varphi_0^0 \wedge \omega^0$ from above, we can differentiate this equation to find

$$
0 = 2(P - Q)\omega^0 \wedge \omega^1 \wedge \omega^2 \wedge \omega^3 \wedge \omega^4.
$$

This implies that $P - Q = 0$, and combined with our normalization $P + Q = 0$, we have

$$
P = Q = 0,
$$

which somewhat simplifies our structure equations (2.15). In particular, we have modulo $\{I\}$

$$
d\varphi_0^0 \equiv (P_3^2 - P_1^4)\omega^1 \wedge \omega^3 - (P_3^1 + P_2^4)\omega^2 \wedge \omega^3 + (P_4^2 + P_1^3)\omega^1 \wedge \omega^4 - (P_4^1 - P_2^3)\omega^2 \wedge \omega^4.
$$

$$
(2.16)
$$

As before, the next step is to study the 8 torsion coefficients P_3^1, P_4^1, P_3^2, P_4^2, P_1^3, P_2^3, P_1^4, P_2^4. We can again obtain a description of how they vary along the connected components of the fibers using infinitesimal methods, and then get a full description of their variation along fibers by explicitly calculating how they transform under one representative of each component of the structure group G_1.

We state only the result of this calculation. The torsion in each fiber transforms by an 8-dimensional linear representation of the group G_1, which decomposes as the direct sum of two 4-dimensional representations. Motivated by (2.16), we define a pair of 2×2 matrix-valued functions on B_1

$$
S_1(u) = \begin{pmatrix} P_3^1 - P_2^4 & P_4^1 + P_2^3 \\ P_3^2 + P_1^4 & P_4^2 - P_1^3 \end{pmatrix}(u), \quad S_2(u) = \begin{pmatrix} P_3^1 + P_2^4 & P_4^1 - P_2^3 \\ P_3^2 - P_1^4 & P_4^2 + P_1^3 \end{pmatrix}(u).
$$

Now, for $g_1 \in G_1$ as in (2.14), one finds that

$$
S_1(u \cdot g_1) = aA^{-1}S_1(u)B, \quad S_2(u \cdot g_1) = aA^{-1}S_2(u)B.
$$

In particular, the two summand representations for our torsion are the same, when restricted to the components of G_1 of (2.14). However, one may also verify that

$$
\begin{aligned}
S_1(u \cdot J) &= -\begin{pmatrix} 0 & 1 \\ -1 & 0 \end{pmatrix} S_1^t(u) \begin{pmatrix} 0 & -1 \\ 1 & 0 \end{pmatrix}, \\
S_2(u \cdot J) &= \begin{pmatrix} 0 & 1 \\ -1 & 0 \end{pmatrix} S_2^t(u) \begin{pmatrix} 0 & -1 \\ 1 & 0 \end{pmatrix}.
\end{aligned}
$$

An immediate conclusion to be drawn from this is that if $S_1(u) = 0$ at some point, then $S_1(u) = 0$ everywhere on the same fiber of $B_1 \to M$, and likewise for S_2.

If the torsion vector takes its values in a union of non-trivial orbits having conjugate stabilizers, then we can try to make a further reduction to the subbundle consisting of those coframes on which the torsion lies in a family of normal forms. However, it is usually interesting in equivalence problems to consider the case when no further reduction is possible; in the present situation, this occurs when all of the invariants vanish identically.

We first claim that $S_2 = 0$ identically if and only if the uniquely determined form φ_0^0 is closed. To see this, note first that from (2.16) we have $S_2 = 0$ if and only if

$$d\varphi_0^0 = \mu \wedge \omega^0$$

for some 1-form μ. We differentiate modulo $\{\omega^0\}$ to obtain

$$0 \equiv -\mu \wedge d\omega^0 \pmod{\{\omega^0\}}$$

which by symplectic linear algebra implies that

$$\mu \equiv 0 \pmod{\{\omega^0\}}.$$

But then $d\varphi_0^0 = 0$, as claimed. Conversely, if $d\varphi_0^0 = 0$, then obviously $S_2 = 0$.

Now suppose that $S_1 = S_2 = 0$ identically. Then because $d\varphi_0^0 = 0$, we can locally find a function $\lambda > 0$ satisfying

$$\varphi_0^0 = \lambda^{-1} d\lambda.$$

We can also compute in case $S_1 = S_2 = 0$ that

$$d(\omega^1 \wedge \omega^2) = -\varphi_0^0 \wedge \omega^1 \wedge \omega^2,$$

so that

$$d(\lambda \omega^1 \wedge \omega^2) = 0.$$

Now, by a variant of the Darboux theorem, this implies that there are locally defined functions p, x such that

$$-dp \wedge dx = \lambda \omega^1 \wedge \omega^2.$$

Similar reasoning gives locally defined functions q, y such that

$$-dq \wedge dy = \lambda \omega^3 \wedge \omega^4.$$

In terms of these functions, note that

$$d(\lambda \omega^0) = \lambda(\omega^1 \wedge \omega^2 + \omega^3 \wedge \omega^4) = -dp \wedge dx - dq \wedge dy,$$

which by the Poincaré lemma implies that there is another locally defined function z such that

$$\lambda \omega^0 = dz - p\,dx - q\,dy.$$

The linear independence of $\omega^0, \ldots, \omega^4$ implies that pulled back by any 1-adapted coframe (that is, any section of B_1), the functions x, y, z, p, q form local coordinates on M. In terms of these local coordinates, our hyperbolic Monge-Ampere system is

$$\begin{aligned} \mathcal{E} &= \{\omega^0, \ \omega^1 \wedge \omega^2 + \omega^3 \wedge \omega^4, \omega^1 \wedge \omega^2 - \omega^3 \wedge \omega^4\} \quad (2.17) \\ &= \{dz - p\,dx - q\,dy, \ dp \wedge dx + dq \wedge dy, \ dp \wedge dx - dq \wedge dy\}. \quad (2.18) \end{aligned}$$

In an obvious way, transverse local integral surfaces of $\mathcal{E}$ are in one-to-one correspondence with solutions to the wave equation for $z(x, y)$

$$\frac{\partial^2 z}{\partial x \, \partial y} = 0.$$

This establishes the following.

Theorem 2.1 *A hyperbolic Monge-Ampere system $(M^5, \mathcal{E})$ satisfies $S_1 = S_2 = 0$ if and only if it is locally equivalent to the Monge-Ampere system (2.18) for the linear homogeneous wave equation.*

This gives us an easily computable method for determining when a given second-order scalar Monge-Ampere equation in two variables is *contact-equivalent* to this wave equation.

Looking at the equation (2.16) for $d\varphi_0^0 \pmod{\{I\}}$, it is natural to ask about the situation in which $S_2 = 0$, but possibly $S_1 \neq 0$. This gives an alternative version of the solution to the inverse problem discussed in the previous chapter.

Theorem 2.2 *A hyperbolic Monge-Ampere system $(M^5, \mathcal{E})$ is locally equivalent to an Euler-Lagrange system if and only if its invariant S_2 vanishes identically.*

Proof. The condition for our $\mathcal{E}$ to contain a Poincaré-Cartan form

$$\Pi = \lambda \, \omega^0 \wedge (\omega^1 \wedge \omega^2 - \omega^3 \wedge \omega^4)$$

is that this Π be *closed* for some function λ on B_1, which we can assume satisfies $\lambda > 0$. Differentiating then gives

$$0 = (d\lambda - 2\lambda\varphi_0^0) \wedge \omega^0 \wedge (\omega^1 \wedge \omega^2 - \omega^3 \wedge \omega^4).$$

Exterior algebra shows that this is equivalent to $d\lambda - 2\lambda\varphi_0^0$ being a multiple of ω^0, say

$$d\lambda - 2\lambda\varphi_0^0 = \sigma \, \lambda \, \omega^0$$

for some function σ, or in other words,

$$d(\log \lambda) - 2\varphi_0^0 = \sigma \, \omega^0.$$

Such an equation can be satisfied if and only if $d\varphi_0^0$ is equivalent modulo $\{I\}$ to a multiple of $d\omega^0$. But we know that

$$d\omega^0 \equiv \omega^1 \wedge \omega^2 + \omega^3 \wedge \omega^4,$$

and from (2.16) we see that $d\varphi_0^0$ is a multiple of this just in case $S_2 = 0$. $\qquad\square$

This result may be thought of as follows. For any hyperbolic Monge-Ampere system, $d\varphi_0^0 \in \Omega^2(B_1)$ is both closed and semibasic for $B_1 \to M$. This means that it is the pullback of a 2-form on M canonically associated with $\mathcal{E}$.[2] We showed that this 2-form vanishes if and only if $S_2 = 0$, which is equivalent to $\mathcal{E}$ being locally Euler-Lagrange. This condition is reminiscent of the vanishing of a curvature, when φ_0^0 is viewed as a connection in the contact line bundle I.

2.2   Neo-Classical Poincaré-Cartan Forms

We now turn to the geometry of Poincaré-Cartan forms in case $n \geq 3$. In the preceding section, we emphasized the corresponding Monge-Ampere system; from now on, we will instead emphasize the more specialized Poincaré-Cartan form.

Let M^{2n+1} be a manifold with contact line bundle I, locally generated by a 1-form θ. Let $\Pi \in \Omega^{n+1}(M)$ be a closed $(n+1)$-form locally expressible as

$$\Pi = \theta \wedge \Psi,$$

where $\bar{\Psi} \in P^n(T^*M/I)$ is primitive modulo $\{I\}$. As in the preceding section, the pointwise linear algebra of this data involves the action of the conformal symplectic group $CSp(n, \mathbf{R})$ on the space $P^n(\mathbf{R}^{2n}) \subset \bigwedge^n \mathbf{R}^{2n}$. When $n = 2$, there are four orbits (including $\{0\}$) for this action, but for $n > 2$, the situation is more complicated. For example, when $n = 3$, the space of primitive 3-forms on $\mathbf{R}^6$ has two open orbits and many degenerate orbits, while for $n = 4$ there are no open orbits.

Which orbits contain the Poincaré-Cartan forms of most interest to us? Consider the classical case, in which $M = J^1(\mathbf{R}^n, \mathbf{R})$, $\theta = dz - p_i dx^i$, and $\Lambda = L(x, z, p)dx$. We have already seen that

$$\begin{align}
\Pi &= d(L\,dx + \theta \wedge L_{p_i} dx_{(i)}) \tag{2.19}\\
&= -\theta \wedge (d(L_{p_i}) \wedge dx_{(i)} - L_z dx) \tag{2.20}\\
&= -\theta \wedge (L_{p_i p_j} dp_j \wedge dx_{(i)} + (L_{p_i z} p_i + L_{p_i x^i} - L_z)dx). \tag{2.21}
\end{align}$$

This suggests the following definition, which singles out Poincaré-Cartan forms of a particular algebraic type; it is these—with a slight refinement in the case $n = 3$, to be introduced below—whose geometry we will study. Note that non-degeneracy of the functional is built in to the definition.

Definition 2.3 *A closed $(n+1)$-form Π on a contact manifold (M^{2n+1}, I) is almost-classical if it can locally be expressed as*

$$\Pi = -\theta \wedge (H^{ij}\pi_i \wedge \omega_{(j)} - K\omega) \tag{2.22}$$

---

[2] This statement also requires one to verify that $d\varphi_0^0$ is invariant under the action of some element of each connected component of G_1; this is easily done.

for some coframing $(\theta, \omega^i, \pi_i)$ of M with $\theta \in \Gamma(I)$, some invertible matrix of functions (H^{ij}), and some function K.

Later, we will see the extent to which this definition generalizes the classical case. We remark that the almost-classical forms $\Pi = \theta \wedge \Psi$ are those for which the primitive $\bar{\Psi}$ lies in the tangent variety of the cone of totally decomposable[3] n-forms in $P^n(T^*M/I)$, but not in the cone itself.

Applying the equivalence method will yield differential invariants and geometric structures intrinsically associated to our Poincaré-Cartan forms. This will be carried out in §2.4, but prior to this, it is best to directly look for some naturally associated geometry. The preview that this provides will make easier the task of interpreting the results of the equivalence method.

First note that the local coframings and functions appearing in the definition of an almost-classical form are not uniquely determined by Π. The extent of the non-uniqueness of the coframings is described in the following lemma, which prepares us for the equivalence method.

Lemma 2.1 *If $(\theta, \omega^i, \pi_i)$ is a coframing adapted to an almost-classical form Π as in Definition 2.3, then $(\bar{\theta}, \bar{\omega}^i, \bar{\pi}_i)$ is another if and only if the transition matrix is of the form (in blocks of size $1, n, n$)*

$$\begin{pmatrix} \bar{\theta} \\ \bar{\omega}^i \\ \bar{\pi}_i \end{pmatrix} = \begin{pmatrix} a & 0 & 0 \\ C^i & A^i_j & 0 \\ D_i & E_{ij} & B^j_i \end{pmatrix} \begin{pmatrix} \theta \\ \omega^j \\ \pi_j \end{pmatrix}.$$

Proof. That the first row of the matrix must be as shown is clear from the requirement that $\theta, \bar{\theta} \in \Gamma(I)$. The real content of the lemma is that Pfaffian system

$$J_\Pi \overset{def}{=} \mathrm{Span}\{\theta, \omega^1, \ldots, \omega^n\}$$

is uniquely determined by Π. This follows from the claim that J_Π is characterized as the set of 1-forms ξ such that $\xi \wedge \Pi$ is totally decomposable; this claim we leave as an exercise for the reader. □

The Pfaffian system $J_\Pi = \{\theta, \omega^1, \ldots, \omega^n\}$ associated to Π is crucial for all that follows. It is canonical in the sense that any local diffeomorphism of M preserving Π also preserves J_Π. In the classical case described previously we have $J_\Pi = \{dz, dx^1, \ldots, dx^n\}$, which is integrable and has leaf space $J^0(\mathbf{R}^n, \mathbf{R})$.

Proposition 2.2 *If $n \geq 4$, then for any almost-classical form Π on a contact manifold (M^{2n+1}, I), the Pfaffian system J_Π is integrable.*

Proof. We need to show that $d\theta, d\omega^i \equiv 0 \pmod{\{J_\Pi\}}$, for some (equivalently, any) coframing $(\theta, \omega^i, \pi_i)$ adapted to Π as in the definition. We write

$$\Pi = -\theta \wedge (H^{ij}\pi_i \wedge \omega_{(j)} - K\omega),$$

---

[3] An n-form is *totally decomposable* if it is equal to the exterior product of n 1-forms.

and
$$d\theta \equiv a^{ij}\pi_i \wedge \pi_j \quad (\mathrm{mod}\ \{J_\Pi\}).$$

Then taking those terms of the equation $d\Pi \equiv 0 \ (\mathrm{mod}\ \{I\})$ that are cubic in π_i, we find
$$a^{ij}\pi_i \wedge \pi_j \wedge H^{kl}\pi_l = 0.$$

Then the 2-form $a^{ij}\pi_i \wedge \pi_j$ has at least $n \geq 3$ linearly independent 1-forms as divisors, which is impossible unless $a^{ij}\pi_i \wedge \pi_j = 0$. Therefore, $d\theta \equiv 0 \ (\mathrm{mod}\ \{J_\Pi\})$ (with only the hypothesis $n \geq 3$).

For the next step, it is useful to work with the 1-forms
$$\pi^i \overset{def}{=} H^{ij}\pi_j,$$

and write
$$d\omega^i \equiv P^i_{jk}\pi^j \wedge \pi^k \quad (\mathrm{mod}\ \{J_\Pi\}), \qquad P^i_{jk} + P^i_{kj} = 0.$$

From the form of Π (2.22), we have
$$0 = \omega^i \wedge \omega^j \wedge \Pi$$

for any pair of indices $1 \leq i, j \leq n$. Differentiating, we obtain
$$\begin{aligned}
0 &= (d\omega^i \wedge \omega^j - \omega^i \wedge d\omega^j) \wedge \Pi \\
&= (P^i_{kl}\pi^k \wedge \pi^l \wedge \pi^j - P^j_{kl}\pi^k \wedge \pi^l \wedge \pi^i) \wedge \theta \wedge \omega \\
&= (\delta^j_m P^i_{kl} - \delta^i_m P^j_{kl})\pi^k \wedge \pi^l \wedge \pi^m \wedge \theta \wedge \omega.
\end{aligned}$$

It is now an exercise in linear algebra to show that if $n \geq 4$, then this implies $P^i_{jk} = 0$. The hypotheses are that $P^i_{jk} = -P^i_{kj}$ and
$$(\delta^j_m P^i_{kl} - \delta^i_m P^j_{kl}) + (\delta^j_k P^i_{lm} - \delta^i_k P^j_{lm}) + (\delta^j_l P^i_{mk} - \delta^i_l P^j_{mk}) = 0. \qquad (2.23)$$

By contracting first on jk and then on il, one finds that for $n \neq 2$ the contraction P^i_{ik} vanishes. Contracting (2.23) only on jk and using $P^i_{ik} = 0$, one finds that for $n \neq 3$, all P^i_{jk} vanish. $\qquad \square$

There do exist counterexamples in case $n = 3$, for which (2.23) implies only that
$$d\omega^i = P^{ij}\pi_{(j)}, \qquad P^{ij} = P^{ji}.$$

For example, if we fix constants $P^{ij} = P^{ji}$ also satisfying $P^{ii} = 0$, then there is a unique simply connected, 7-dimensional Lie group G having a basis of left-invariant 1-forms $(\omega^i, \theta, \pi^i)$ satisfying structure equations
$$d\omega^i = P^{ij}\pi_{(j)}, \qquad d\theta = -\pi^i \wedge \omega^i, \qquad d\pi^i = 0.$$

In this case, θ generates a homogeneous contact structure on G, and the form
$$\Pi \overset{def}{=} -\theta \wedge \pi^i \wedge \omega_{(i)}$$

is closed, giving an almost-classical form for which J_Π is not integrable.

These counterexamples cannot arise from classical cases, however, and this suggests that we consider the following narrower class of Poincaré-Cartan forms.

Definition 2.4 *An almost-classical Poincaré-Cartan form Π is neo-classical if its associated Pfaffian system J_Π is integrable.*

So the preceding Proposition states that in case $n \geq 4$, every almost-classical Poincaré-Cartan form is neo-classical, and we have narrowed the definition only in case $n = 3$.

The foliation corresponding to the integrable Pfaffian system J_Π is the beginning of the very rich geometry associated to a neo-classical Poincaré-Cartan form. Before investigating it further, we justify the study of this class of objects with the following.

Proposition 2.3 *Every neo-classical Poincaré-Cartan form Π on a contact manifold (M, I) is locally equivalent to that arising from some classical variational problem. More precisely, given such (M, I, Π), there are local coordinates (x^i, z, p_i) on M with respect to which the contact system I is generated by $dz - p_i dx^i$, and there is a Lagrangian of the form $L(x^i, z, p_i)dx$ whose Poincaré-Cartan form is Π.*

Note that we have already observed the converse, that those non-degenerate Poincaré-Cartan forms arising form classical variational problems (in case $n \geq 3$) are neo-classical.

Proof. We fix a coframing $(\theta, \omega^i, \pi_i)$ as in the definition of an almost-classical form. Using the Frobenius theorem, we take independent functions (x^i, z) on M so that

$$J_\Pi = \{\omega^i, \theta\} = \{dx^i, dz\}.$$

By relabeling if necessary, we may assume $\theta \notin \{dx^i\}$, and we find that there are functions p_i so that

$$\theta \in \mathbf{R} \cdot (dz - p_i dx^i).$$

The fact that $\theta \wedge (d\theta)^n \neq 0$ implies that (x^i, z, p_i) are local coordinates on M.

We now introduce a technical device that is often useful in the study of exterior differential systems. Let

$$\mathcal{F}^p \Omega^q \subset \Omega^{p+q}(M)$$

be the collection of $(p+q)$-forms with at least p factors in J_Π; this is well-defined. With this notation, the fact that J_Π is integrable may be expressed as

$$d(\mathcal{F}^p \Omega^q) \subset \mathcal{F}^p \Omega^{q+1}.$$

There is a version of the Poincaré lemma that can be applied to each leaf of the foliation determined by J_Π, with smooth dependence on the leaves' parameters; it says precisely that the complex

$$\mathcal{F}^p \Omega^0 \xrightarrow{d} \mathcal{F}^p \Omega^1 \xrightarrow{d} \cdots$$

is locally exact for each p. Now, any almost-classical form Π lies in $\mathcal{F}^n\Omega^1$; so not only is the closed form Π locally equal to $d\Lambda$ for some $\Lambda \in \Omega^n(M)$, we can actually choose Λ to lie in $\mathcal{F}^n\Omega^0$. In other words, we can locally find a Lagrangian Λ of the form

$$\Lambda = L^0(x, z, p)dx + L^i(x, z, p)dz \wedge dx_{(i)}$$

for some functions L^0, L^i. This may be rewritten as

$$\Lambda = (L^0 + p_iL^i)dx + \theta \wedge (L^idx_{(i)}),$$

and then the condition $\theta \wedge d\Lambda = 0$ (recall that this was part of the construction of the Poincaré-Cartan form associated to any class in $H^n(\Omega^*/\mathcal{I})$) gives the relation

$$L^i(x, z, p) = \frac{\partial L}{\partial p_i}(x, z, p).$$

This is exactly the condition for Λ to locally be a classical Lagrangian.   □

Returning to the geometry associated to a neo-classical Poincaré-Cartan form Π, we have found (or in case $n = 3$, postulated) an integrable Pfaffian system J_Π which is invariant under contact transformations preserving Π. Locally in M, the induced foliation has a smooth "leaf-space" Q of dimension $n + 1$, and there is a smooth submersion $q : M \to Q$ whose fibers are n-dimensional integral manifolds of J_Π. On such a neighborhood, the foliation will be called *simple*, and as we are only going to consider the local geometry of Π in this section, we assume that the foliation is simple on all of M. We may restrict to smaller neighborhoods as needed in the following.

To explore the geometry of the situation, we ask what the data (M^{2n+1}, I, Π) look like from the point of view of Q^{n+1}. The first observation is that we can locally identify M, as a contact manifold, with the standard contact manifold $G_n(TQ)$, the Grassmannian bundle parameterizing n-dimensional subspaces of fibers of TQ. This is easily seen in coordinates as follows. If, as in the preceding proof, we integrate J_Π as

$$J_\Pi = \{dz, dx^i\}$$

for some local functions z, x^i on M, then the same functions z, x^i may be regarded as coordinates on Q. With the assumption that $\theta \notin \{dx^i\}$ (on M, again), we must have $dz - p_idx^i \in \Gamma(I)$ for some local functions p_i on M, which by the non-degeneracy condition for I make (x^i, z, p_i) local coordinates on M. These p_i can also thought of as local fiber coordinates for $M \to Q$, and we can map $M \to G_n(TQ)$ by

$$(x^i, z, p_i) \mapsto ((x^i, z); \{dz - p_idx^i\}^\perp).$$

The latter notation refers to a hyperplane in the tangent space of Q at (x^i, z). Under this map, the standard contact system on $G_n(TQ)$ evidently pulls back to I, so we have a local contact diffeomorphism commuting with projections to

Q. Every point transformation of Q prolongs to give a contact transformation of $G_n(TQ)$, hence of M as well. Conversely, every contact transformation of M that preserves Π is the prolongation of a point transformation of Q, because the foliation by integral manifolds of J_Π defining Q is associated to Π in a contact-invariant manner.[4] In this sense, studying the geometry of a neo-classical Poincaré-Cartan form (in case $n \geq 3$) under contact transformations is locally no different from studying the geometry of an equivalence class of classical non-degenerate first-order scalar Lagrangians under point transformations.

We have now interpreted (M, I) as a natural object in terms of Q, but our real interest lies in Π. What kind of geometry does Π define in terms of Q? We will answer this question in terms of the following notion.

Definition 2.5 *A Lagrangian potential for a neo-classical Poincaré-Cartan form* Π *on M is an n-form* $\Lambda \in \mathcal{F}^n \Omega^0$ *(that is, Λ is semibasic for $M \to Q$) such that $d\Lambda = \Pi$.*

We saw in the proof of Proposition 2.3 that locally a Lagrangian potential Λ exists. Such Λ are not unique, but are determined only up to addition of closed forms in $\mathcal{F}^n \Omega^0$. It will be important below to note that a closed form in $\mathcal{F}^n \Omega^0$ must actually be basic for $M \to Q$; that is, it must be locally the pullback of a (closed) n-form on Q. In particular, the difference between any two Lagrangian potentials for a give neo-classical form Π must be basic.

Consider one such Lagrangian potential Λ, semibasic over Q. Then at each point $m \in M$, one may regard Λ_m as an element of $\bigwedge^n (T^*_{q(m)} Q)$, an n-form at the corresponding point of Q. This defines a map

$$\nu : M \to \bigwedge^n (T^* Q),$$

commuting with the natural projections to Q. Counting dimensions shows that if ν is an immersion, then we actually obtain a hypersurface in $\bigwedge^n (T^* Q)$; to be more precise, we have a smoothly varying field of hypersurfaces in the vector bundle $\bigwedge^n (T^* Q) \to Q$. It is not hard to see that ν is an immersion if the Poincaré-Cartan form Π is non-degenerate, which is a standing hypothesis. We can work backwards as well: given a hypersurface $M \hookrightarrow \bigwedge^n (T^* Q)$ over an $(n+1)$-dimensional manifold Q, we may restrict to M the tautological n-form on $\bigwedge^n (T^* Q)$ to obtain a form $\Lambda \in \Omega^n(M)$. Under mild technical hypotheses on the hypersurface M, the form $d\Lambda \in \Omega^{n+1}(M)$ will be a neo-classical Poincaré-Cartan form.

So we have associated to a Poincaré-Cartan form Π, and a choice of Lagrangian potential $\Lambda \in \mathcal{F}^n \Omega^0$, a field of hypersurfaces in $\bigwedge^n (T^* Q) \to Q$. However, we noted that Λ was not canonically defined in terms of Π, so neither are these hypersurfaces. As we have seen, the ambiguity in Λ is that another admissible $\tilde{\Lambda}$ may differ from Λ by a form that is basic over Q. This means that $\Lambda - \tilde{\Lambda}$ does not depend on the fiber coordinate for $M \to Q$, and therefore

---

[4]This statement is valid only in case the foliation by integral manifolds of J_Π is simple; in other cases, only a cumbersome local version of the statement holds.

the two corresponding immersions $\nu, \tilde{\nu}$ differ in each fiber M_q $(q \in Q)$ only by a translation in $\bigwedge^n(T_q^*Q)$. Consequently, we have in each $\bigwedge^n(T_q^*Q)$ a hypersurface well-defined up to translation. A contact transformation of M which preserves Π will therefore carry the field of hypersurfaces for a particular choice of Λ to a field of hypersurfaces differing by (a field of) *affine transformations*.

To summarize,

> *one can canonically associate to any neo-classical Poincaré-Cartan form (M, Π) a field of hypersurfaces in the bundle $\bigwedge^n(T^*Q) \to Q$, regarded as a bundle of affine spaces. We expect the differential invariants of Π to include information about the geometry of each of these affine hypersurfaces, and this will turn out to be the case.*

2.3   Digression on Affine Geometry of Hypersurfaces

Let $\mathbf{A}^{n+1}$ denote $(n+1)$-dimensional affine space, which is simply $\mathbf{R}^{n+1}$ regarded as a homogeneous space of the group $A(n+1)$ of affine transformations

$$x \mapsto g \cdot x + v, \qquad g \in GL(n+1, \mathbf{R}), \ v \in \mathbf{R}^{n+1}.$$

Let $x : \mathbf{F} \to \mathbf{A}^{n+1}$ denote the principal $GL(n+1, \mathbf{R})$-bundle of affine frames; that is,

$$\mathbf{F} = \{f = (x, (e_0, \dots, e_n))\},$$

where $x \in \mathbf{A}^{n+1}$ is a point, and $(e_0, \dots, e_n)$ is a basis for the tangent space $T_x\mathbf{A}^{n+1}$. The action is given by

$$(x, (e_0, \dots, e_n)) \cdot (g_b^a) \overset{def}{=} (x, (e_b g_0^b, \dots, e_b g_n^b)). \tag{2.24}$$

For this section, we adopt the index ranges $0 \leq a, b, c \leq n$, $1 \leq i, j, k \leq n$, and always assume $n \geq 2$.

There is a basis of 1-forms ω^a, φ_b^a on $\mathbf{F}$ defined by decomposing the $\mathbf{A}^{n+1}$-valued 1-forms

$$dx = e_a \cdot \omega^a, \quad de_a = e_b \cdot \varphi_a^b.$$

These equations implicitly use a trivialization of $T\mathbf{A}^{n+1}$ that commutes with affine transformations. Differentiating, we obtain the structure equations for $\mathbf{F}$:

$$d\omega^a = -\varphi_b^a \wedge \omega^b, \quad d\varphi_b^a = -\varphi_c^a \wedge \varphi_b^c. \tag{2.25}$$

Choosing a reference frame $f_0 \in \mathbf{F}$ determines an identification $\mathbf{F} \cong A(n+1)$, and under this identification the 1-forms ω^a, φ_b^a on $\mathbf{F}$ correspond to a basis of left-invariant 1-forms on the Lie group $A(n+1)$. The structure equations (2.25) on $\mathbf{F}$ then correspond to the usual Maurer-Cartan structure equations for left-invariant 1-forms on a Lie group.

In this section, we will study the geometry of smooth hypersurfaces $M^n \subset \mathbf{A}^{n+1}$, to be called *affine hypersurfaces*, using the method of moving frames; no previous knowledge of this method is assumed. In particular, we give constructions that associate to M geometric objects in a manner invariant under affine transformations of the ambient $\mathbf{A}^{n+1}$. Among these objects are tensor fields H_{ij}, U^{ij}, and T_{ijk} on M, called the affine first and second fundamental forms and the affine cubic form of the hypersurface. We will classify those non-degenerate (to be defined) hypersurfaces for which $T_{ijk} = 0$ everywhere. This is of interest because the particular neo-classical Poincaré-Cartan forms that we study later induce fields of affine hypersurfaces of this type.

Suppose given a smooth affine hypersurface $M \subset \mathbf{A}^{n+1}$. We define the collection of 0-*adapted frames* along M by

$$\mathbf{F}_0(M) = \{(x, (e_0, \ldots, e_n)) \in \mathbf{F} : x \in M,\ e_1, \ldots, e_n \text{ span } T_x M\} \subset \mathbf{F}.$$

This is a principal subbundle of $\mathbf{F}|_M$ whose structure group is[5]

$$G_0 \overset{def}{=} \left\{ g_0 = \begin{pmatrix} a & 0 \\ v & A \end{pmatrix} : a \in \mathbf{R}^*,\ A \in GL(n, \mathbf{R}),\ v \in \mathbf{R}^n \right\}. \tag{2.26}$$

Restricting forms on $\mathbf{F}$ to $\mathbf{F}_0(M)$ (but supressing notation), we have

$$\omega^0 = 0, \qquad \omega^1 \wedge \cdots \wedge \omega^n \neq 0.$$

Differentiating the first of these gives

$$0 = d\omega^0 = -\varphi_i^0 \wedge \omega^i,$$

and we apply the Cartan lemma to obtain

$$\varphi_i^0 = H_{ij}\omega^j \text{ for some functions } H_{ij} = H_{ji}.$$

One way to understand the meaning of these functions H_{ij}, which constitute the *first fundamental form* of $M \subset \mathbf{A}^{n+1}$, is as follows. At any given point of $M \subset \mathbf{A}^{n+1}$, one can find an affine frame and associated coordinates with respect to which M is locally a graph

$$x^0 = \tfrac{1}{2}\bar{H}_{ij}(x^1, \ldots, x^n)x^i x^j$$

for some functions $\bar{H}_{ij}$. Restricted to the 0-adapted frame field defined by

$$\bar{e}_0(x) = \frac{\partial}{\partial x^0}, \quad \bar{e}_i(x) = (\bar{H}_{ij}(x)x^j + \tfrac{1}{2}\partial_i \bar{H}_{jk}(x)x^j x^k)\frac{\partial}{\partial x^0} + \frac{\partial}{\partial x^i},$$

one finds that the values over $0 \in M$ of the functions H_{ij} equal $\bar{H}_{ij}(0)$. Loosely speaking, the functions H_{ij} express the second derivatives of a defining function for M.

---

[5] Here and throughout, $\mathbf{R}^*$ denotes the connected group of positive real numbers under multiplication.

Returning to the general situation, we calculate as follows. We substitute the expression $\varphi_i^0 = H_{ij}\omega^j$ into the structure equation $d\varphi_i^0 = -\varphi_b^0 \wedge \varphi_i^b$, collect terms, and conclude

$$0 = (dH_{ij} + H_{ij}\varphi_0^0 - H_{kj}\varphi_i^k - H_{ik}\varphi_j^k) \wedge \omega^j.$$

Using the Cartan lemma, we have

$$dH_{ij} = -H_{ij}\varphi_0^0 + H_{kj}\varphi_i^k + H_{ik}\varphi_j^k + T_{ijk}\omega^k$$

for some functions $T_{ijk} = T_{ikj} = T_{kji}$. This infinitesimally describes how the functions H_{ij} vary along the fibers of $\mathbf{F}_0(M)$, on which $\omega^j = 0$. In particular, as a matrix-valued function $H = (H_{ij})$ on $\mathbf{F}_0(M)$, it transforms by a linear representation of the structure group:

$$H(f \cdot g_0) = (a^{-1})\,{}^tAH(f)A,$$

where $g_0 \in G_0$ is as in (2.26).[6] Now we consider the quantity

$$\Delta(f) \overset{def}{=} \det(H_{ij}(f)),$$

which vanishes at some point of $\mathbf{F}_0(M)$ if and only if it vanishes on the entire fiber containing that point. We will say that $M \subset \mathbf{A}^{n+1}$ is *non-degenerate* if $\Delta \neq 0$ everywhere on $\mathbf{F}_0(M)$. Also note that the absolute signature of H_{ij} is well-defined at each point of M. It is easy to see that H_{ij} is definite if and only if $M \subset \mathbf{A}^{n+1}$ is convex. In what follows, we will assume that M is a non-degenerate hypersurface, but not necessarily that it is convex.

It turns out that $T = (T_{ijk})$, which one would like to regard as a sort of covariant derivative of $H = (H_{ij})$, is not a tensor; that is, it does not transform by a linear representation along the fibers of $\mathbf{F}_0(M) \to M$. We will exploit this below to reduce the principal bundle $\mathbf{F}_0(M) \to M$ to a subbundle of frames satisfying a higher-order adaptivity condition. Namely, $\mathbf{F}_1(M) \subset \mathbf{F}_0(M)$ will consist of those frames where T_{ijk} is traceless with respect to the non-degenerate symmetric bilinear form H_{ij}, meaning $H^{jk}T_{ijk} = 0$, where (H^{ij}) is the matrix inverse of (H_{ij}). Geometrically, the reduction will amount to a canonical choice of line field $\mathbf{R}e_0$ transverse to M, which we will think of as giving at each point of M a canonical *affine normal line*.

To justify this, we let (H^{ij}) denote the matrix inverse of (H_{ij}), and let

$$C_i \overset{def}{=} H^{jk}T_{ijk}$$

be the vector of traces of T with respect to H. We compute

$$\begin{aligned}
d(\log \Delta) &= \Delta^{-1}d\Delta \\
&= \mathrm{Tr}(H^{-1}dH) \\
&= H^{ij}dH_{ij} \\
&= -n\varphi_0^0 + 2\varphi_i^i + C_i\omega^i.
\end{aligned}$$

---

[6] As usual, our argument only proves this claim for g_0 in the identity component of G_0, but it may be checked directly for representative elements of each of the other components.

Now differentiate again and collect terms to find

$$0 = (dC_i - C_j\varphi_i^j - (n+2)H_{ij}\varphi_0^j) \wedge \omega^i. \qquad (2.27)$$

Therefore, we have

$$dC_i \equiv C_j\varphi_i^j + (n+2)H_{ij}\varphi_0^j \pmod{\{\omega^1,\ldots,\omega^n\}}, \qquad (2.28)$$

which expresses how the traces C_i vary along the fibers of $\mathbf{F}_0(M) \to M$. In particular, if the matrix (H_{ij}) is non-singular, as we are assuming, then the action of the structure group on the values of the vector $(C_i) \in \mathbf{R}^n$ is transitive; that is, every value in $\mathbf{R}^n$ is taken by (C_i) in each fiber. Therefore, the set of 0-adapted frames $f \in \mathbf{F}_0(M)$ where each $C_i(f) = 0$ is a principal subbundle $\mathbf{F}_1(M) \subset \mathbf{F}_0(M)$, whose structure group is the stabilizer of $0 \in \mathbf{R}^n$ under the action. This stabilizer is

$$G_1 \overset{def}{=} \left\{ g_1 = \begin{pmatrix} a & 0 \\ 0 & A \end{pmatrix} : a \in \mathbf{R}^*,\ A \in GL(n,\mathbf{R}) \right\}.$$

Comparing to the full action (2.24) of the affine group $A(n+1)$ on $\mathbf{F}$, we see that along each fiber of $\mathbf{F}_1(M)$, the direction $\mathbf{R}e_0$ is fixed. Thus, we have uniquely chosen the direction of e_0 at each point of M by the condition $C_i = 0$ for $i = 1,\ldots,n$.

A more concrete explanation of what we have done is seen by locally presenting our hypersurface in the form

$$x^0 = \frac{1}{2}\bar{H}_{ij}(0)x^i x^j + \frac{1}{6}\bar{T}_{ijk}(x^1,\ldots,x^n)x^i x^j x^k.$$

An affine change of coordinates that will preserve this form is the addition of a multiple of x^0 to each x^i; the n choices that this entails can be uniquely made so that $\bar{T}_{ijk}(0)$ is traceless with respect to $\bar{H}_{ij}(0)$. Once such choices are fixed, then so is the direction of $\frac{\partial}{\partial x^0}$, and this gives the canonical affine normal line at $x = 0$.

There is a remarkable interpretation of the affine normal direction at a point where H_{ij} is positive-definite (see [Bla67]). Consider the 1-parameter family of hyperplanes parallel to the tangent plane at the given point. For those planes sufficiently near the tangent plane, the intersection with a fixed neighborhood in the surface is a closed submanifold of dimension $n-2$ in M, having an affine-invariant center-of-mass. These centers-of-mass form a curve in affine space, passing through the point of interest; this curve's tangent line at that point is the affine normal direction.

We can see from (2.28) that on $\mathbf{F}_1(M)$, where T_{ijk} is traceless, the forms φ_0^j are semibasic over M. It is less convenient to express these in terms of the basis ω^i than to express them in terms of $\varphi_i^0 = H_{ij}\omega^j$, assuming that $M \subset \mathbf{A}^{n+1}$ is non-degenerate. On $\mathbf{F}_1(M)$ we write

$$\varphi_0^j = U^{jk}\varphi_k^0.$$

Now (2.27), restricted to $\mathbf{F}_1(M)$ where $C_i = 0$, implies that $U^{ij} = U^{ji}$.

The reader may carry out computations similar to those above to show that the (U^{ij}) and (T_{ijk}) are tensors; that is, they transform along the fibers of $\mathbf{F}_1(M)$ by a linear representation of G_1. For example, if $T_{ijk} = 0$ at some point of $\mathbf{F}_1(M)$, then $T_{ijk} = 0$ everywhere along the same fiber of $\mathbf{F}_1(M) \to M$. Furthermore, the transformation law for U^{ij} is such that if $U^{ij} = \lambda H^{ij}$ at some point, for some λ, then the same is true—with possibly varying λ—everywhere on the same fiber. We can now give some additional interpretations of the simplest cases of the affine second fundamental form U^{ij} and the affine cubic form T_{ijk}. The following theorem is the main purpose of this digression.

Theorem 2.3 *(1) If $U^{ij} = \lambda H^{ij}$ everywhere on $\mathbf{F}_1(M)$—that is, if the second fundamental form is a scalar multiple of the first fundamental form—then either $\lambda = 0$ everywhere or $\lambda \neq 0$ everywhere. In the first case, the affine normal lines of M are all parallel, and in the second case, the affine normal lines of M are all concurrent.*

(2) If $T_{ijk} = 0$ everywhere on $\mathbf{F}_1(M)$, then $U^{ij} = \lambda H^{ij}$ everywhere. In this case, if $\lambda = 0$, then M is a paraboloid, while if $\lambda \neq 0$, then M is a non-degenerate quadric.

Proof. Suppose first that $U^{ij} = \lambda H^{ij}$ on $\mathbf{F}_1(M)$ for some function λ. This is same as writing

$$\varphi_0^j = \lambda H^{jl} \varphi_l^0 = \lambda \omega^j.$$

We differentiate this equation (substituting itself), and obtain

$$(d\lambda - \lambda \varphi_0^0) \wedge \omega^j = 0 \text{ for each } j.$$

Under the standing assumption $n > 1$, this means that

$$d\lambda = \lambda \varphi_0^0.$$

So assuming that M is connected, we have the first statement of (1). We will describe the geometric consequences of each of the two possibilities.

First, suppose that $\lambda = 0$, so that $U^{ij} = 0$, and then $\varphi_0^j = 0$ throughout $\mathbf{F}_1(M)$. Then the definition of our original basis of 1-forms gives

$$de_0 = e_a \varphi_0^a = e_0 \varphi_0^0,$$

meaning that the direction in $\mathbf{A}^{n+1}$ of e_0 is fixed throughout $\mathbf{F}_1(M)$, or equivalently, all of the affine normals of M are parallel.

Next, suppose $\lambda \neq 0$, and assume for simplicity that $\lambda < 0$. The differential equation $d\lambda = \lambda \varphi_0^0$ implies that we can restrict to the principal subbundle $\mathbf{F}_2(M)$ where $\lambda = -1$. This amounts to a choice of a particular vector field e_0 along the affine normal line field already defined. Note that on $\mathbf{F}_2(M)$, we have

$$\varphi_0^j = -\omega^j = -H^{jk} \varphi_k^0, \qquad \varphi_0^0 = 0. \tag{2.29}$$

As a result, the structure equations $dx = e_i \omega^i$ and $de_0 = e_a \varphi_0^a = -e_i \omega^i$ imply that

$$d(x + e_0) = 0,$$

so that $x + e_0$ is a constant element of $\mathbf{A}^{n+1}$. In particular, all of the affine normal lines of M pass through this point. This completes the proof of (1).

Now assume that $T_{ijk} = 0$ identically; this will be satisfied by each member of the fields of affine hypersurfaces associated to certain neo-classical Poincaré-Cartan forms of interest. Our first claim is that $U^{ij} = \lambda H^{ij}$ for some function λ on $\mathbf{F}_1(M)$. To see this, note that our hypothesis means

$$dH_{ij} = -H_{ij}\varphi_0^0 + H_{kj}\varphi_i^k + H_{ik}\varphi_j^k.$$

We differentiate this, using the structure equations in the simplified form that defined the reduction to $\mathbf{F}_1(M)$, and obtain

$$0 = -H_{kj}\varphi_0^k \wedge \varphi_i^0 - H_{ik}\varphi_0^k \wedge \varphi_j^0.$$

If we use H_{ij} to raise and lower indices and define

$$U_{ij} = H_{ik}H_{lj}U^{kl},$$

then the preceding equation may be written as

$$0 = -(U_{jl}H_{ik} + U_{il}H_{jk})\omega^l \wedge \omega^k.$$

The coefficients of this vanishing 2-form then satisfy

$$0 = U_{jl}H_{ik} + U_{il}H_{jk} - U_{jk}H_{il} - U_{ik}H_{jl};$$

we multiply by H^{ik} (and sum over i, k) to conclude

$$U_{jl} = \frac{1}{n}(H^{ik}U_{ik})H_{jl}.$$

This proves that

$$U^{ij} = \lambda H^{ij},$$

with $\lambda = \frac{1}{n}H^{kl}U_{kl}$.

We now return to the possibilities $\lambda = 0$, $\lambda \neq 0$ under the stronger hypothesis $T_{ijk} = 0$.

In the first case, note that with the condition $\varphi_j^i = 0$ on $\mathbf{F}_1(M)$, we have that the Pfaffian system generated by φ_0^0 and φ_j^i (for $1 \leq i, j \leq n$) is integrable. Let $\tilde{M}$ be any leaf of this system. Restricted to $\tilde{M}$, we have

$$dH_{ij} = 0,$$

so that the functions H_{ij} are constants. Furthermore, the linearly independent 1-forms ω^i on $\tilde{M}$ are each closed, so that (at least locally, or else on a simply connected cover) there are coordinates u^i on $\tilde{M}$ with

$$\omega^i = du^i.$$

Substituting all of this into the structure equations, we have:

- $de_0 = 0$, so that e_0 is a constant element of $\mathbf{A}^{n+1}$ on $\tilde{M}$;

- $de_i = e_0\varphi_i^0 = e_0 H_{ij}\omega^j = d(e_0 H_{ij}u^j)$, so that

$$e_i = \bar{e}_i + e_0 H_{ij}u^j$$

  for some constant $\bar{e}_i \in \mathbf{A}^{n+1}$;

- $dx = e_i\omega^i = (\bar{e}_i + e_0 H_{ij}u^j)du^i = d(u^i\bar{e}_i + \frac{1}{2}e_0 H_{ij}u^iu^j)$, so that

$$x = \bar{x} + u^i\bar{e}_i + \frac{1}{2}H_{ij}u^iu^je_0$$

  for some constant $\bar{x} \in \mathbf{A}^{n+1}$.

The conclusion is that as the coordinates u^i vary on $\tilde{M}$, the $\mathbf{A}^{n+1}$-valued function x on $\tilde{M}$ traces out a paraboloid, with vertex at $\bar{x}$ and axis along the direction of e_0.

Turning to the case $\lambda \neq 0$, recall that under the assumption $\lambda < 0$, we can reduce to a subbundle $\mathbf{F}_2(M) \subset \mathbf{F}_1(M)$ on which $\lambda = -1$. We use the differential equation

$$dH_{ij} = H_{ik}\varphi_j^k + H_{kj}\varphi_i^k$$

to reduce again to a subbundle $\mathbf{F}_3(M) \subset \mathbf{F}_2(M)$ on which $H_{ij} = \bar{H}_{ij}$ is some constant matrix. On $\mathbf{F}_3(M)$, the forms φ_j^i satisfy linear algebraic relations

$$0 = \bar{H}_{ik}\varphi_j^k + \bar{H}_{kj}\varphi_i^k.$$

Our assumption $\lambda = -1$ allows us to combine these with the relations (2.29) by defining

$$\Phi = \begin{pmatrix} 0 & \varphi_j^0 \\ \varphi_0^i & \varphi_j^i \end{pmatrix}, \qquad \mathbf{H} = \begin{pmatrix} 1 & 0 \\ 0 & \bar{H} \end{pmatrix},$$

and then

$$\mathbf{H}\Phi + {}^t\Phi\mathbf{H} = 0.$$

In other words, the matrix-valued 1-form Φ on $\mathbf{F}_3(M)$ takes values in the Lie algebra of the stabilizer of the bilinear form $\mathbf{H}$. For instance, if our hypersurface M is convex, so that (H_{ij}) is definite everywhere, then we could have chosen $\bar{H}_{ij} = \delta_{ij}$, and then Φ would take values in the Lie algebra $\mathfrak{so}(n+1,\mathbf{R})$. Whatever the signature of H_{ij}, let the stabilizer of $\mathbf{H}$ be denoted by $O(\mathbf{H}) \subset GL(n+1,\mathbf{R})$, with Lie algebra $\mathfrak{so}(\mathbf{H})$. Then the structure equation

$$d\Phi + \Phi \wedge \Phi = 0$$

implies that there is locally (alternatively, on a simply connected cover) a map

$$g : \mathbf{F}_3(M) \to O(\mathbf{H})$$

such that

$$\Phi = g^{-1}dg.$$

Using the structure equations $de_a = e_b \varphi_a^b$, this implies

$$d(e_a \cdot (g^{-1})_b^a) = 0,$$

so that

$$e_a = \bar{e}_b g_a^b$$

for some fixed affine frame $(\bar{e}_b)$. In particular, the $\mathbf{A}^{n+1}$-valued function e_0 on $\mathbf{F}_3(M)$ takes as its values precisely the points of a level surface of a non-degenerate quadratic form, defined by $\mathbf{H}$. Recalling from the first part of the proof that $x + e_0$ is constant on $\mathbf{A}^{n+1}$, this means that the hypersurface M, thought of as the image of the map $x : \mathbf{F}_3(M) \to \mathbf{A}^{n+1}$, is a constant translate of a non-degenerate quadric hypersurface. The signature of the quadric is (p, q), where $(p - 1, q)$ is the signature of the first fundamental form (H_{ij}).

The case $\lambda > 0$ instead of $\lambda < 0$ is quite similar, but M is a quadric of signature (p, q) when (H_{ij}) has signature $(p, q - 1)$. $\qquad \square$

2.4 The Equivalence Problem for $n \geq 3$

We now consider a contact manifold (M, I) with a closed, almost-classical form

$$\Pi = -\theta \wedge (H^{ij} \pi_i \wedge \omega_{(j)} - K\omega). \tag{2.30}$$

We will shortly specialize to the case in which Π is neo-classical. The coframes in which Π takes the form (2.30), for some functions H^{ij} and K, constitute a G-structure as described in Lemma 2.1. The purpose of this section is to describe a canonical reduction of this G-structure to one carrying a pseudo-connection satisfying structure equations of a prescribed form, as summarized in (2.47–2.48), at least in case the matrix (H^{ij}) is either positive- or negative-definite everywhere. This application of the equivalence method involves no techniques beyond those introduced in §2.1, but some of the linear-algebraic computations are more involved.

We begin by refining our initial G-structure as follows.

Lemma 2.2 *Let (M, I) be a contact manifold with almost-classical form Π. (1) There exist local coframings $(\theta, \omega^i, \pi_i)$ on M such that Π has the form (2.30) and such that*

$$d\theta \equiv -\pi_i \wedge \omega^i \pmod{\{I\}}.$$

(2) Local coframings as in (1) are the sections of a G_0-structure $B_0 \to M$, where G_0 is the group of matrices of the form (in blocks of size $1, n, n$)

$$g_0 = \begin{pmatrix} a & 0 & 0 \\ C^i & A_j^i & 0 \\ D_i & S_{ik} A_j^k & a(A^{-1})_i^j \end{pmatrix}, \quad A \in GL(n, \mathbf{R}), \ S_{ij} = S_{ji}. \tag{2.31}$$

(3) If two local coframings as in (1) are related as

$$\begin{pmatrix} \theta \\ \omega^i \\ \pi_i \end{pmatrix} = g_0^{-1} \cdot \begin{pmatrix} \bar{\theta} \\ \bar{\omega}^j \\ \bar{\pi}_j \end{pmatrix},$$

and if $\Pi = -\theta \wedge (H^{ij} \pi_i \wedge \omega_{(j)} - K\omega) = -\bar{\theta} \wedge (\bar{H}^{ij} \bar{\pi}_i \wedge \bar{\omega}_{(j)} - \bar{K}\bar{\omega})$ *are the expressions for* Π *with respect to these coframings, then*

$$\begin{align}
H &= a^2 (\det A) A^{-1} \bar{H} \, {}^t A^{-1}, \tag{2.32}\\
K &= a(\det A)(\bar{K} - Tr(\bar{H}S)). \tag{2.33}
\end{align}$$

Proof. (1) First observe that in any coframing, we may write

$$d\theta \equiv a^{ij} \pi_i \wedge \pi_j + b_i^j \pi_j \wedge \omega^i + c_{ij} \omega^i \wedge \omega^j \pmod{\{I\}}.$$

We will deal with each of the three coefficient matrices (a^{ij}), (b_i^j), (c_{ij}) to obtain the desired condition $d\theta \equiv -\sum \pi_i \wedge \omega^i$.

- The proof of Proposition 2.2 showed for $n \geq 3$ that

$$0 \equiv d\theta \equiv a^{ij} \pi_i \wedge \pi_j \pmod{\{J_\Pi\}},$$

  which implies $a^{ij} \pi_i \wedge \pi_j = 0$. This followed from calculating $0 = d\Pi$ modulo $\{I\}$.

- From the fact that θ is a contact form, we have

$$0 \neq \theta \wedge (d\theta)^n = \pm \det(b_i^j) \theta \wedge \omega \wedge \pi,$$

  so that (b_i^j) is an invertible matrix. Therefore, we may apply the matrix $-(b_i^j)^{-1}$ to the 1-forms π_j to obtain a new basis in which we have $b_i^j = -\delta_i^j$, so that

$$d\theta \equiv -\pi_i \wedge \omega^i + c_{ij} \omega^i \wedge \omega^j \pmod{\{I\}}.$$

  Note that this coframe change is of the type admitted by Lemma 2.1, preserving the form (2.30).

- Finally, we can replace π_i by $\pi_i + c_{ij} \omega^j$ to have the desired $d\theta \equiv -\pi_i \wedge \omega^i$. This coframe change also preserves the form (2.30).

(2) We already know that any matrix as in Lemma 2.1 will preserve the form (2.30). We write the action of such a matrix as

$$\begin{cases}
\bar{\theta} &= a\theta \\
\bar{\omega}^i &= C^i \theta + A_j^i \omega^j \\
\bar{\pi}_i &= D_i \theta + S_{ik} A_j^k \omega^j + B_i^j \pi_j.
\end{cases}$$

It is easily verified that the condition $d\theta \equiv -\pi_i \wedge \omega^i$ implies the analogous condition $d\bar{\theta} \equiv -\bar{\pi}_i \wedge \bar{\omega}^i$ if and only if

$$\begin{cases} B_i^j A_k^i = a\delta_k^j, \\ S_{jk} = S_{kj}. \end{cases}$$

This is what we wanted to prove.

(3) These formulae are seen by substituting the formulae for $(\bar{\theta}, \bar{\omega}^i, \bar{\pi}_i)$ into the equation for the two expressions for Π, and comparing terms. One uses the following fact from linear algebra: if

$$\bar{\omega}^i \equiv A_j^i \omega^j \pmod{\{I\}},$$

then

$$\bar{\omega}_{(j)} \equiv (\det A)(A^{-1})_j^i \omega_{(i)} \pmod{\{I\}};$$

that is, the coefficients of $\bar{\omega}_{(j)}$ in terms of $\omega_{(i)}$ are the cofactors of the coefficient matrix of $\bar{\omega}^i$ in terms of ω^j. $\qquad\square$

We can see from (2.32) that the matrix $H = (H^{ij})$ transforms under coframe changes like a bilinear form, up to scaling, and in particular that its absolute signature is fixed at each point of M. To proceed, we have to assume that this signature is constant throughout M. In particular, we shall from now on assume that H is positive or negative definite everywhere, and refer to almost-classical forms Π with this property as *definite*. Cases of different constant signature are of interest, but can be easily reconstructed by the reader in analogy with the definite case examined below.

Once we assume that the matrix-valued function H on B_0 is definite, the following is an easy consequence of the preceding lemma.

Lemma 2.3 *Given a definite, almost-classical Poincaré-Cartan form Π on a contact manifold (M, I), there are 0-adapted local coframings $(\theta, \omega^i, \pi_i)$ for which*

$$\Pi = -\theta \wedge (\delta^{ij} \pi_i \wedge \omega_{(j)}),$$

and these form a G_1-structure $B_1 \subset B_0 \to M$, where G_1 is the group of matrices g_1 of the form (2.31) with

$$\det A > 0, \quad a(\det A)^{\frac{1}{2}} A \in O(n, \mathbf{R}), \quad S_{ii} = 0.$$

This follows from imposing the conditions $\bar{H} = H = I_n$, $\bar{K} = K = 0$ in the previous lemma. Unfortunately, it is difficult to give a general expression in coordinates for such a 1-adapted coframing in the classical case, because such an expression requires that we normalize the Hessian matrix $(L_{p_i p_j})$. In practice, however, such a coframing is usually easy to compute.

It is convenient for later purposes to use a different parameterization of our group G_1. Namely, an arbitrary element will be written as

$$g_1 = \begin{pmatrix} \pm r^{n-2} & 0 & 0 \\ C^i & r^{-2}A^i_j & 0 \\ D_i & S_{ik}A^k_j & \pm r^n(A^{-1})^j_i \end{pmatrix}, \qquad (2.34)$$

where $A = (A^i_j) \in SO(n, \mathbf{R})$, $r > 0$, $S_{ij} = S_{ji}$, $S_{ii} = 0$. Also, now that the orthogonal group has appeared, some of the representations occurring in the sequel are isomorphic to their duals, for which it may be unuseful and sometimes confusing to maintain the usual summation convention, in which one only contracts a pair of indices in which one index is raised and the other lowered. Therefore, we will now sum any index occuring twice in a single term, regardless of its positions.

We now assume that we have a definite, neo-classical Poincaré-Cartan form Π with associated G_1-structure $B_1 \to M$, and we begin searching for differential invariants. There are local pseudo-connection 1-forms $\rho, \gamma^i, \delta_i, \alpha^i_j, \sigma_{ij}$ defined so that equations of the following form hold:

$$d\begin{pmatrix} \theta \\ \omega^i \\ \pi_i \end{pmatrix} = -\begin{pmatrix} (n-2)\rho & 0 & 0 \\ \gamma^i & -2\rho\delta^i_j + \alpha^i_j & 0 \\ \delta_i & \sigma_{ij} & n\rho\delta^j_i - \alpha^j_i \end{pmatrix} \wedge \begin{pmatrix} \theta \\ \omega^j \\ \pi_j \end{pmatrix} + \begin{pmatrix} \Theta \\ \Omega^i \\ \Pi_i \end{pmatrix},$$

where θ, ω^i, π_i are the tautological 1-forms on B_1, the torsion 2-forms Θ, Ω^i, Π_i are semibasic for $B_1 \to M$, and the pseudo-connection 1-forms satisfy

$$\alpha^i_j + \alpha^j_i = 0, \qquad \sigma_{ij} = \sigma_{ji}, \qquad \sigma_{ii} = 0.$$

These last conditions mean that the pseudo-connection matrix takes values in the Lie algebra $\mathfrak{g}_1 \subset \mathfrak{gl}(2n+1, \mathbf{R})$ of G_1.

The pseudo-connection 1-forms are not uniquely determined, and our next step is to exploit this indeterminacy to try to absorb components of the torsion.

First, we know that $d\theta \equiv -\pi_i \wedge \omega^i \pmod{\{I\}}$. The difference between $\Theta = d\theta + (n-2)\rho \wedge \theta$ and $-\pi_i \wedge \omega^i$ is therefore a semibasic multiple of θ, which can be absorbed by a semibasic change in ρ. We can therefore simply assume that

$$d\theta = -(n-2)\rho \wedge \theta - \pi_i \wedge \omega^i,$$

or equivalently, $\Theta = -\pi_i \wedge \omega^i$.

Second, our assumption that Π is neo-classical means that the Pfaffian system $J_\Pi = \{\theta, \omega^i\}$ is integrable (even up on B_1). In the structure equation

$$d\omega^i = -\gamma^i \wedge \theta - (-2\rho\delta^i_j + \alpha^i_j) \wedge \omega^j + \Omega^i, \qquad (2.35)$$

this means that $\Omega^i \equiv 0 \pmod{\{J_\Pi\}}$. Also, Ω^i is semibasic over M, so we can write

$$\Omega^i \equiv T^{ijk}\pi_j \wedge \omega^k + \frac{1}{2}P^i_{jk}\omega^j \wedge \omega^k \pmod{\{I\}}. \qquad (2.36)$$

Now, adding semibasic 1-forms to γ^i allows us to preserve the equation (2.35) while also making (2.36) an equality, and not merely a congruence. A little linear algebra shows that there is a *unique* linear combination of the ω^i that can be added to α^i_j, preserving $\alpha^i_j + \alpha^j_i = 0$, to absorb the term $\frac{1}{2} P^i_{jk} \omega^j \wedge \omega^k$. This leaves us only with

$$\Omega^i = T^{ijk} \pi_j \wedge \omega^k.$$

As in the elimination of the P^i_{jk}, we can add a combination of the π_i to α^i_j to arrange

$$T^{ijk} = T^{kji}.$$

To investigate the third torsion term Π_i, we use an alternate derivation of the equation for $d\pi_i$. Namely, we differentiate the equation

$$d\theta = -(n-2)\rho \wedge \theta - \pi_i \wedge \omega^i,$$

and take the result only modulo $\{I\}$ to avoid the unknown quantity $d\rho$. This eventually yields

$$0 \equiv -(\Pi_k - T^{ijk} \pi_i \wedge \pi_j) \wedge \omega^k \quad (\mathrm{mod}\ \{I\}).$$

As before, multiples of θ may be absorbed by redefining δ_i, so that we can assume this congruence is an equality. Reasoning similar to that which proves the Cartan lemma gives

$$\Pi_k - T^{ijk} \pi_i \wedge \pi_j = \nu_{kl} \wedge \omega^l$$

for some semibasic 1-forms $\nu_{kl} = \nu_{lk}$. Now, most of these forms ν_{kl} can be subtracted from the pseudo-connection forms σ_{kl}, simplifying the torsion; but the condition $\sigma_{ii} = 0$ prevents us from completely absorbing them. Instead, the trace remains, and we have

$$\Pi_k = \delta_{kl} \nu \wedge \omega^l + T^{ijk} \pi_i \wedge \pi_j.$$

We can learn more about ν using the integrability condition $d\Pi = 0$, taken modulo terms quadratic in the π_i:

$$0 = d\Pi \equiv n\theta \wedge \nu \wedge \omega.$$

A consequence is that $\nu \equiv 0 \pmod{\{\theta, \omega^i\}}$; in other words, ν has no π_i-terms, and may be written (using again a change in δ_i) as

$$\nu = \sum N_i \omega^i.$$

Then replacing σ_{ij} by

$$\sigma_{ij} + \tfrac{n}{n+2}(\delta_{ik} N_j + \delta_{jk} N_i - \tfrac{2}{n}\delta_{ij} N_k)\omega^k$$

yields new pseudo-connection forms, for which the third torsion term is simply

$$\Pi_k = T^{ijk} \pi_i \wedge \pi_j.$$

This completes the major step of absorbing torsion by altering the pseudo-connection.

Before proceeding to the next major step, we look for linear-algebraic conditions on the torsion which may simplify later calculations. In particular, we made only very coarse use of $d\Pi = 0$ above. Now we compute more carefully

$$0 = d\Pi = -\theta \wedge (2T^{ijk} + \delta^{ij}T^{lkl})\pi_k \wedge \pi_j \wedge \omega_{(i)},$$

so we must have

$$2T^{ijk} + \delta^{ij}T^{lkl} = 2T^{ikj} + \delta^{ik}T^{ljl}. \qquad (2.37)$$

The next major step is a reduction of our G_1-structure. We will examine the variation of the functions $T^j \stackrel{def}{=} T^{iji}$ along fibers of $B_1 \to M$, and observe that the zero-locus $\{T^j = 0\}$ defines a G_2-structure for a certain codimension-n subgroup $G_2 \subset G_1$.

As usual, the variation of T^j will be described infinitesimally. To study dT^j without knowledge of the traceless part of dT^{ijk}, we exploit the exterior algebra, writing

$$d(\theta \wedge \omega^1 \wedge \cdots \wedge \omega^n) = ((n+2)\rho + T^k\pi_k) \wedge \theta \wedge \omega. \qquad (2.38)$$

We will differentiate this for information about dT^k, but in doing so we will need information about $d\rho$ as well. Fortunately, this is available by differentiating the first structure equation

$$d\theta = -(n-2)\rho \wedge \theta - \pi_k \wedge \omega^k,$$

yielding

$$(n-2)d\rho \equiv \gamma^k \wedge \pi_k \quad (\mathrm{mod} \ \{\theta, \omega^i\}).$$

Now we return to differentiating (2.38) and eventually find

$$dT^k \equiv -\tfrac{n+2}{n-2}\gamma^k + (n\delta_j^k\rho - \alpha_j^k)T^j \quad (\mathrm{mod} \ \{\theta, \omega^i, \pi_i\}).$$

This means that along fibers of $B_1 \to M$, the vector-valued function $T(u) = (T^j(u))$, $u \in B_1$, is orthogonally rotated (infinitesimally, by α_j^k), scaled (by ρ), and translated (by γ^i). In fact, for $g_1 \in G_1$ as in (2.34),

$$T(u \cdot g_1) = \pm r^2 A^{-1}(r^{n-2}T(u) - \tfrac{n+2}{n-2}C).$$

Now the set

$$B_2 \stackrel{def}{=} \{u \in B_1 : T(u) = 0\} \subset B_1$$

is a G_2-subbundle of $B_1 \to M$, where G_2 consists of matrices as in (2.34) with $T^i = 0$.

On the submanifold $B_2 \subset B_1$, we have from (2.37) the symmetry

$$T^{ijk} = T^{kji} = T^{ikj}.$$

As a consequence, the torsion Π_k restricts to

$$\Pi_k = T^{ijk}\pi_j \wedge \pi_k = 0.$$

The previous structure equations continue to hold, but the forms $\gamma^i|_{B_2}$ should not be regarded as part of the pseudo-connection, as they are now semibasic over M. We therefore write

$$d\begin{pmatrix} \theta \\ \omega^i \\ \pi_i \end{pmatrix} = -\begin{pmatrix} (n-2)\rho & 0 & 0 \\ 0 & -2\rho\delta^i_j + \alpha^i_j & 0 \\ \delta_i & \sigma_{ij} & n\rho\delta^j_i - \alpha^j_i \end{pmatrix} \wedge \begin{pmatrix} \theta \\ \omega^j \\ \pi_j \end{pmatrix} + \begin{pmatrix} \Theta \\ \Omega^i \\ \Pi_i \end{pmatrix},$$
(2.39)

where still

$$\alpha^i_j + \alpha^j_i = 0, \ \sigma_{ij} = \sigma_{ji}, \ \sigma_{ii} = 0,$$
(2.40)

and now

$$\begin{cases} \Theta = -\pi_i \wedge \omega^i, \\ \Omega^i = -(S^i_j\omega^j + U^{ij}\pi_j) \wedge \theta + T^{ijk}\pi_j \wedge \omega^k, \\ \Pi_i = 0. \end{cases}$$
(2.41)

Here we have denoted $\gamma^i \equiv S^i_j\omega^j + U^{ij}\pi_j \pmod{\{I\}}$. Also, we still have

$$T^{ijk} = T^{kji} = T^{ikj}, \quad T^{iik} = 0.$$
(2.42)

Notice that we can alter α^i_j and ρ to assume that

$$S^i_j = S^j_i, \qquad S^i_i = 0,$$
(2.43)

where we also have to add combinations of ω^i to δ_i to preserve $\Pi_i = 0$. In fact, these assumptions uniquely determine α^i_j and ρ, although δ_i and σ_{ij} still admit some ambiguity.

Equations (2.39–2.43) summarize the results of the equivalence method carried out to this point. We have uncovered the primary differential invariants of a definite neo-classical Poincaré-Cartan form: they are the functions T^{ijk}, S^i_j and U^{ij}. Their properties are central in what follows.

For example, note that the rank-n Pfaffian system $\{\omega^i\}$ on B_2 is invariant under the action of the structure group G_2, and therefore it is the pullback of a Pfaffian system (also to be denoted $\{\omega^i\}$) down on M. Testing its integrability, we find

$$d\omega^i \equiv -U^{ij}\pi_j \wedge \theta \pmod{\{\omega^i\}}.$$
(2.44)

We will see shortly that the matrix-valued function (U^{ij}) varies along the fibers of $B_2 \to M$ by a linear representation of G_2, so that it is plausible to ask about those Poincaré-Cartan forms for which $U^{ij} = 0$; (2.44) shows that this is equivalent to the integrability of $\{\omega^i\}$. In this case, in addition to the local fibration $M \to Q$ whose fibers are leaves of J_Π, we have

$$Q^{n+1} \to N^n,$$

where N is the locally defined n-dimensional "leaf space" for $\{\omega^i\}$. Coordinates on N—equivalently, functions on M whose differentials lie in $\{\omega^i\}$—may be thought of as "preferred independent variables" for the contact-equivalence class of our Euler-Lagrange equation, canonical in the sense that every symmetry of M preserving the Poincaré-Cartan form preserves the fibration $M \to N$ and therefore acts on N. Note that even if an (M, Π) satisfying $U^{ij} = 0$ came to us from a classical Lagrangian with independent variables (x^i), we need not have $\{\omega^i\} = \{dx^i\}$.

This is not to say that the case $U^{ij} \neq 0$ is uninteresting. In the next section, we will see an important family of examples from Riemannian geometry with $U^{ij} = \lambda \delta^i_j$. To obtain preliminary information about U^{ij} in a manner that will not require much knowledge of S^i_j or T^{ijk}, we start with the equation

$$d(\omega^1 \wedge \cdots \wedge \omega^n) = 2n\rho \wedge (\omega^1 \wedge \cdots \wedge \omega^n) + U^{ij}\theta \wedge \pi_j \wedge \omega_{(i)}. \tag{2.45}$$

We will differentiate again, but we need more refined information about $d\rho$; this is obtained from

$$0 = d^2\theta = -((n-2)d\rho + \delta_i \wedge \omega^i + \pi_i \wedge \gamma^i) \wedge \theta.$$

Keep in mind that $\gamma^i = S^i_j\omega^j + U^{ij}\pi_j \pmod{\{I\}}$ on this reduced bundle. We can now write

$$(n-2)d\rho + \delta_i \wedge \omega^i + \pi_i \wedge \gamma^i = \tau \wedge \theta \tag{2.46}$$

for some unknown 1-form τ. Returning to the derivative of (2.45), we find

$$0 \equiv \tfrac{2n}{n-2}(-\pi_i \wedge U^{ij}\pi_j) \wedge \omega + U^{ij}\pi_i \wedge \pi_j \wedge \omega \pmod{\{I\}}.$$

This implies that $U^{ij}\pi_i \wedge \pi_j \wedge \omega = 0$, so that we have

$$U^{ij} = U^{ji}.$$

We will need an even more refined version of the equation (2.46) for $d\rho$. In the preceding paragraph, we substituted that equation into the equation for $0 \equiv d^2(\omega^1 \wedge \cdots \wedge \omega^n) \pmod{\{I\}}$. Now, we substitute it instead into

$$\begin{aligned}
0 &\equiv d^2(\omega^1 \wedge \cdots \wedge \omega^n) \pmod{\{\pi_1, \ldots, \pi_n\}} \\
&\equiv \left(\tfrac{2n}{n-2}\tau - U^{ij}\sigma_{ij}\right) \wedge \theta \wedge \omega^1 \wedge \cdots \wedge \omega^n.
\end{aligned}$$

This means that $\tfrac{2n}{n-2}\tau - U^{ij}\sigma_{ij}$ lies in $\{\theta, \omega^i, \pi_i\}$. Recall that also $\gamma^i = S^i_j\omega^j + U^{ij}\pi_j + V^i\theta$ for some functions V^i, and we can put this back into (2.46) to finally obtain

$$(n-2)d\rho = -\delta_i \wedge \omega^i - S^i_j\pi_i \wedge \omega^j + \left(\tfrac{n-2}{2n}\right)U^{ij}\sigma_{ij} \wedge \theta + (s_i\omega^i - t^i\pi_i) \wedge \theta,$$

for some functions s_i, t^i. Furthermore, we can replace each δ_i by $\delta_i - s_i\theta$, preserving previous equations, to assume that $s_i = 0$. This gives

$$(n-2)d\rho = -\delta_i \wedge \omega^i - S^i_j\pi_i \wedge \omega^j + \left(\tfrac{n-2}{2n}\right)U^{ij}\sigma_{ij} \wedge \theta - t^i\pi_i \wedge \theta,$$

which will be used in later sections.

The last formulae that we will need are those for the transformation rules for T^{ijk}, U^{ij}, S^i_j along fibers of $B \to M$. These are obtained by computations quite similar to those carried out above, and we only state the results here, which are:

- $dT^{ijk} \equiv n\rho T^{ijk} - \alpha_l^i T^{ljk} - \alpha_l^j T^{ilk} - \alpha_l^k T^{ijl},$

- $dU^{ij} \equiv 2n\rho U^{ij} - \alpha_l^i U^{lj} - \alpha_l^j U^{il},$

- $dS_j^i \equiv (n-2)\rho S_j^i - \alpha_l^i S_j^l + S_l^i \alpha_j^l + \frac{1}{2}(U^{il}\sigma_{lj} + U^{jl}\sigma_{li}) - \frac{1}{n}\delta_j^i U^{kl}\sigma_{kl} + T^{ijk}\delta_k,$

all modulo $\{\theta, \omega^i, \pi_i\}$. Notice in particular that T^{ijk} and U^{ij} transform by a combination of rescaling and a standard representation of $SO(n)$. However, (S_j^i) is only a tensor when the tensors (T^{ijk}) and (U^{ij}) both vanish. We will consider this situation in the next chapter.

An interpretation of the first two transformation rules is that the objects

$$\mathbf{T} = T^{ijk}(\pi_i \circ \pi_j \circ \pi_k) \otimes |\pi_1 \wedge \cdots \wedge \pi_n|^{-\frac{2}{n}},$$
$$\mathbf{U} = U^{ij}\pi_i \circ \pi_j$$

are invariant modulo $J_\Pi = \{\theta, \omega^i\}$ under flows along fibers over M; that is, when restricted to a fiber of $B_2 \to Q$, they actually descend to well-defined objects on the smaller fiber of $M \to Q$. The restriction to fibers suggests our next result, which nicely relates the differential invariants of the Poincaré-Cartan form with the affine geometry of hypersurfaces discussed in the preceding section.

Theorem 2.4 *The functions T^{ijk} and U^{ij} are coefficients of the affine cubic form and affine second fundamental form for the fiberwise affine hypersurfaces in $\bigwedge^n(T^*Q)$ induced by a semibasic Lagrangian potential Λ of Π.*

Proving this is a matter of identifying the bundles where the two sets of invariants are defined, and unwinding the definitions.

In the next section, we will briefly build on the preceding results in the case where $T^{ijk} = 0$ and $U^{ij} \neq 0$, showing that these conditions roughly characterize those definite neo-classical Poincaré-Cartan forms appearing in the problem of finding prescribed mean curvature hypersurfaces, in Riemannian or Lorentzian manifolds. In the next chapter, we will extensively consider the case $T^{ijk} = 0$, $U^{ij} = 0$, which includes remarkable Poincaré-Cartan forms arising in conformal geometry. About the case for which $T^{ijk} \neq 0$, nothing is known.

For reference, we summarize the results of the equivalence method that will be used below. Associated to a definite, neo-classical Poincaré-Cartan form Π on a contact manifold (M, I) is a G-structure $B \to M$, where

$$G = \left\{ \begin{pmatrix} \pm r^{n-2} & 0 & 0 \\ 0 & r^{-2}A_j^i & 0 \\ D_i & S_{ik}A_j^k & \pm r^n(A^{-1})_i^j \end{pmatrix} : \begin{array}{l} (A_j^i) \in SO(n, \mathbf{R}), \ r > 0, \\ S_{ij} = S_{ji}, \ S_{ii} = 0 \end{array} \right\}.$$

$$(2.47)$$

$B \to M$ supports a pseudo-connection (not uniquely determined)

$$\varphi = - \begin{pmatrix} (n-2)\rho & 0 & 0 \\ 0 & -2\rho\delta_j^i + \alpha_j^i & 0 \\ \delta_i & \sigma_{ij} & n\rho\delta_i^j - \alpha_i^j \end{pmatrix},$$

with $\alpha^i_j + \alpha^j_i = 0$, $\sigma_{ij} = \sigma_{ji}$, $\sigma_{ii} = 0$, such that in the structure equation

$$d \begin{pmatrix} \theta \\ \omega^i \\ \pi_i \end{pmatrix} = -\varphi \wedge \begin{pmatrix} \theta \\ \omega^j \\ \pi_j \end{pmatrix} + \tau,$$

the torsion is of the form

$$\tau = \begin{pmatrix} -\pi_i \wedge \omega^i \\ -(S^i_j \omega^j + U^{ij}\pi_j) \wedge \theta + T^{ijk}\pi_j \wedge \omega^k \\ 0 \end{pmatrix},$$

with

$$T^{ijk} = T^{jik} = T^{kji}, \ T^{iik} = 0; \ U^{ij} = U^{ji}; \ S^i_j = S^j_i, \ S^i_i = 0.$$

In terms of any section of $B \to M$, the Poincaré-Cartan form is

$$\Pi = -\theta \wedge \pi_i \wedge \omega_{(i)}.$$

One further structure equation is

$$(n-2)d\rho = -\delta_i \wedge \omega^i - S^i_j \pi_i \wedge \omega^j + \left(\frac{n-2}{2n}\right) U^{ij}\sigma_{ij} \wedge \theta - t^i \pi_i \wedge \theta. \qquad (2.48)$$

2.5   The Prescribed Mean Curvature System

In this section, we will give an application of the part of the equivalence method completed so far. We will show that a definite, neo-classical Poincaré-Cartan form with $T^{ijk} = 0$, and satisfying an additional open condition specified below, is locally equivalent to that which arises in the problem of finding in a given Riemannian manifold a hypersurface whose mean curvature coincides with a prescribed background function. This conclusion is presented as Theorem 2.5.

     To obtain this result, we continue applying the equivalence method where we left off in the preceding section, and take up the case $T^{ijk} = 0$. From our calculations in affine hypersurface geometry, we know that this implies that

$$U^{ij} = \lambda \delta^{ij},$$

for some function λ on the principal bundle $B \to M$; alternatively, this can be shown by computations continuing those of the preceding section. We will show that under the hypothesis $\lambda < 0$, the Poincaré-Cartan form Π is locally equivalent to that occurring in a prescribed mean curvature system.

     We have in general on B that

$$dU^{ij} \equiv 2n\rho U^{ij} - \alpha^i_k U^{kj} - \alpha^j_k U^{ik} \pmod{\{\theta, \omega^i, \pi_i\}}.$$

Then for our $U^{ij} = \lambda \delta^{ij}$, the function λ scales positively along fibers of $B \to M$, so under our assumption $\lambda < 0$ we may make a reduction to

$$B_1 = \{u \in B : \lambda(u) = -1\} \subset B;$$

this defines a subbundle of B of codimension 1, on which ρ is semibasic over M.[7] In particular, on B_1 we may write

$$\rho = -\frac{H}{2n}\theta + E_i\omega^i + F^i\pi_i$$

for some functions H, E_i, F^i. The reason for the normalization of the θ-coefficient will appear shortly.

We claim that $F^i = 0$. To see this, start from the equation (2.48) for $d\rho$, which on B_1 reads

$$(n-2)d\rho = -\delta_i \wedge \omega^i - t^i\pi_i \wedge \theta - S^i_j\pi_i \wedge \omega^j.$$

Then, as we have done so often, we compute $d^2\omega$, where $\omega = \omega^1 \wedge \cdots \wedge \omega^n$ and

$$d\omega^i = 2\rho \wedge \omega^i - \alpha^i_j \wedge \omega^j + \pi_i \wedge \theta - S^i_j\omega^j \wedge \theta.$$

We find

$$d\omega = 2n\rho \wedge \omega - \theta \wedge \pi_i \wedge \omega_{(i)} = 2n\rho \wedge \omega + \Pi,$$

and the next step is simplified by knowing $d\Pi = 0$:

$$
\begin{aligned}
0 &= d^2\omega \\
&= 2n\, d\rho \wedge \omega - 2n\rho \wedge d\omega \\
&= -\left(\frac{2n}{n-2}\right)(t^i\pi_i \wedge \theta) \wedge \omega \\
&\quad + 2n(E_j\omega^j + F^j\pi_j) \wedge \theta \wedge \pi_i \wedge \omega_{(i)} \\
&= 2n\theta \wedge \left(F^j\pi_i \wedge \pi_j \wedge \omega_{(i)} + \left(\frac{t^i}{n-2} + E^i\right)\pi_i \wedge \omega\right).
\end{aligned}
$$

This gives our claim $F^i = 0$, as well as

$$(n-2)E^i = -t^i.$$

For our next reduction, we will show that we can define a principal subbundle

$$B_2 = \{u \in B_1 : E^i(u) = 0\} \subset B_1,$$

having structure group defined by the condition $D_i = 0$, $r = 1$ in (2.47). This follows by computing modulo $\bigwedge^2\{\theta, \omega^i, \pi_i\}$:

$$(n-2)d\rho \equiv -\delta_i \wedge \omega^i,$$

and also

$$d\rho \equiv -\frac{1}{2n}dH \wedge \theta + dE_i \wedge \omega^i - E_j\alpha^j_i \wedge \omega^i.$$

---

[7]In this section, we will denote by B_1, B_2, etc., successive reductions of the G-structure $B \to M$ which was constructed in the preceding section. These are *not* the same as the bundles of the same name used in constructing B, which are no longer needed.

Comparing these, we obtain

$$\left(dE_i - E_j\alpha_i^j + \tfrac{1}{n-2}\delta_i\right) \wedge \omega^i - \tfrac{1}{2n}dH \wedge \theta \equiv 0.$$

This implies that

$$dE_i - E_j\alpha_i^j + \tfrac{1}{n-2}\delta_i \equiv 0 \quad (\text{mod } \{\theta, \omega^i, \pi_i\}),$$

justifying the described reduction to $B_2 \to M$, on which ρ and δ_i are semibasic. Finally, a third reduction is made possible by the general equation

$$dS_j^i \equiv (n-2)\rho S_j^i - \alpha_k^i S_j^k + S_k^i \alpha_j^k + \tfrac{1}{2}(U^{il}\sigma_{lj} + U^{jl}\sigma_{li}) - \tfrac{1}{n}\delta_j^i U^{kl}\sigma_{kl},$$

modulo $\{\theta, \omega^i, \pi_i\}$. On B_2, where in particular $\lambda = -1$ and ρ is semibasic, we have

$$dS_j^i \equiv -\alpha_k^i S_j^k + S_k^i \alpha_j^k - \sigma_{ij} \quad (\text{mod } \{\theta, \omega^i, \pi_i\}).$$

This means that the torsion matrix (S_j^i) can undergo translation by an arbitrary traceless symmetric matrix along the fibers of $B_2 \to M$, so the locus

$$B_3 = \{u \in B_2 : S_j^i(u) = 0\} \subset B_2$$

is a subbundle, whose structure group is $SO(n, \mathbf{R})$ with Lie algebra represented by matrices of the form

$$a_2 = \begin{pmatrix} 0 & 0 & 0 \\ 0 & \alpha_j^i & 0 \\ 0 & 0 & -\alpha_i^j \end{pmatrix}, \qquad \alpha_j^i + \alpha_i^j = 0.$$

This is all the reduction that we shall need. On B_3, we have equations

$$\begin{cases} \rho = -\tfrac{H}{2n}\theta, \\ d\theta = -\pi_i \wedge \omega^i \quad (\text{because } \rho \wedge \theta = 0 \text{ on } B_3), \\ (n-2)d\rho = -\delta_i \wedge \omega^i \quad (\text{because } t^i = -(n-2)E^i = 0 \text{ on } B_3). \end{cases}$$

The δ_i appearing the third equation are semibasic over M, and the three equations together imply that

$$dH \equiv 0 \quad (\text{mod } \{\theta, \omega^i\}).$$

This last observation is quite important. Recall the integrable Pfaffian system $J_{\Pi} = \{\theta, \omega^i\}$, assumed to have a well-defined leaf-space Q^{n+1} with submersion $M \to Q$. The last equation shows that H is locally constant along the fibers of $M \to Q$, and may therefore be thought of as a function on Q.

Now, considering the two structure equations

$$\begin{cases} d\theta = -\pi_i \wedge \omega^i, \\ d\omega^i = 2\rho \wedge \omega^i - \alpha_j^i \wedge \omega^j + \pi_i \wedge \theta, \end{cases}$$

it is tempting to define

$$\tilde{\pi}_i = \pi_i + \frac{H}{n}\omega^i,$$

and rewrite them as

$$d\begin{pmatrix} \theta \\ \omega^i \end{pmatrix} = -\begin{pmatrix} 0 & \tilde{\pi}_j \\ -\tilde{\pi}_i & \alpha^i_j \end{pmatrix} \wedge \begin{pmatrix} \theta \\ \omega^j \end{pmatrix}.$$

Observe that this looks exactly like the structure equation characterizing the Levi-Civita connection of a Riemannian metric. We justify and use this as follows.

Consider the quadratic form on B_3

$$\theta^2 + \sum(\omega^i)^2.$$

An easy computation shows that for any vertical vector field $v \in Ker(\pi_*)$ for $\pi : B_3 \to Q$,

$$\mathcal{L}_v\left(\theta^2 + \sum(\omega^i)^2\right) = 0.$$

This means that our quadratic form is the pullback of a quadratic form on Q, which defines there a Riemannian metric ds^2. There is locally a bundle isomorphism over Q

$$B_3 \to \mathcal{F}(Q, ds^2)$$

from B_3, which was constructed from the neo-classical Poincaré-Cartan form Π, to the orthonormal frame bundle of this Riemannian metric. Under this isomorphism, the Q-semibasic forms θ, ω^i correspond to the tautological semibasic forms on $\mathcal{F}(Q, ds^2)$, while the matrix

$$\begin{pmatrix} 0 & \tilde{\pi}_j \\ -\tilde{\pi}_i & \alpha^i_j \end{pmatrix}$$

corresponds to the Levi-Civita connection matrix. The contact manifold M, as a quotient of B_3, may be then identified with the manifold of tangent hyperplanes to Q; and the Poincaré-Cartan form is

$$\begin{aligned} \Pi &= -\theta \wedge (\pi_i \wedge \omega_{(i)}) \\ &= -\theta \wedge (\tilde{\pi}_i \wedge \omega_{(i)} - H\omega). \end{aligned}$$

We recognize this as exactly the Poincaré-Cartan form for the prescribed mean curvature $H = H(q)$ system, in an arbitrary $(n+1)$-dimensional Riemannian manifold. The following is what we have shown.

Theorem 2.5 *A definite neo-classical Poincaré-Cartan form (M, Π) whose differential invariants satisfy $T^{ijk} = 0$ and $U^{ij} = \lambda\delta^i_j$ with $\lambda < 0$ is locally equivalent to the Poincaré-Cartan of the prescribed mean curvature system on some Riemannian manifold (Q^{n+1}, ds^2).*

We will consider these Poincaré-Cartan forms further in §4.1, when we discuss the formula for the second variation of a Lagrangian functional $\mathcal{F}_\Lambda$. At that time, we will also see an interpretation of the partial reduction $B_2 \supset B_3$ in terms of the Riemannian geometry. Note that it is easy, given (M, Π) as in the proposition, to determine the prescribed function $H(q)$ by carrying out the reductions described above, and to determine the Riemann curvature of the ambient $(n + 1)$-manifold in terms of the connection 1-forms $\bar{\pi}_i$, α_j^i. The Euclidean minimal surface system discussed in §1.4 is the case $H = 0$, $R_{ijkl} = 0$.

The fact that such an (M, Π) canonically determines (Q, ds^2) implies the following.[8]

Corollary 2.1 *The symmetry group of (M, Π) is equal to the group of isometries of (Q, ds^2) that preserve the function H.*

A consequence of this is the fact, claimed in §1.4, that all symmetries of the minimal surface Poincaré-Cartan form—and hence, all classical conservation laws for the Euler-Lagrange equation—are induced by Euclidean motions.

Finally, in case $T^{ijk} = 0$ and $U^{ij} = \lambda \delta_j^i$ with $\lambda > 0$ instead of $\lambda < 0$, one can carry out similar reductions, eventually producing on the quotient space Q^{n+1} a Lorentz metric $ds^2 = -\theta^2 + \sum(\omega^i)^2$; the Poincaré-Cartan form is then equivalent to that for prescribed mean curvature of space-like hypersurfaces.

---

[8] As usual, this assumes that the foliation associated to J_Π is simple; otherwise, only a local reformulation holds.

Chapter 3

Conformally Invariant Euler-Lagrange Systems

Among non-linear Euler-Lagrange equations on $\mathbf{R}^n$, the largest symmetry group that seems to occur is the $\frac{(n+1)(n+2)}{2}$-dimensional conformal group. This consists of diffeomorphisms of the n-sphere that preserve its standard conformal structure, represented by the Euclidean metric under stereographic projection to $\mathbf{R}^n$. These maximally symmetric equations have a number of special properties, including of course an abundance of classical conservation laws as predicted by Noether's theorem. This chapter concerns the geometry of the Poincaré-Cartan forms associated to these equations, and that of the corresponding conservation laws.

We will begin by presenting background material on conformal geometry. This includes a discussion of the flat conformal structure on the n-sphere and its symmetry group, a construction of a canonical parallelized principal bundle over a manifold with conformal structure, and the definition of the *conformal Laplacian*, a second-order differential operator associated to a conformal structure. This material will provide the framework for understanding the geometry of non-linear Poisson equations, in particular the maximally symmetric non-linear example

$$\Delta u = Cu^{\frac{n+2}{n-2}}, \qquad C \neq 0.$$

After developing the geometric context for this equation, we will continue the equivalence problem for Poincaré-Cartan forms, pursuing the branch in which these Euler-Lagrange equations occur.

We then turn to conservation laws for these conformally invariant equations. The elaborate geometric structure allows several approaches to computing these conservation laws, and we will carry out one of them in detail. The analogous development for non-linear wave equations involves conformal structures with Lorentz signature, and the conserved quantities for maximally symmetric Euler-Lagrange equations in this case give rise to integral identities that have been very useful in analysis.

3.1 Background Material on Conformal Geometry

In this section, we discuss some of the less widely known aspects of conformal geometry. In the first subsection, we define a flat model for conformal geometry which is characterized by its large symmetry group, and we give structure equations in terms of the Maurer-Cartan form of this group. In the second subsection, we give Cartan's solution to the local equivalence problem for general conformal structures on manifolds. This consists of an algorithm by which one associates to any conformal structure $(N, [ds^2])$ a parallelized principal bundle $P \to N$ having structure equations of a specific algebraic form. In the third subsection, we introduce a second-order differential operator Δ, called the *conformal Laplacian*, which is associated to any conformal structure and which appears in the Euler-Lagrange equations of conformal geometry that we study in the remainder of the chapter. The fundamental definition is the following.

Definition 3.1 *A* conformal inner-product *at a point* $p \in N$ *is an equivalence class of positive inner-products on* T_pN, *where two such inner-products are equivalent if one is a positive scalar multiple of the other. A* conformal structure, *or* conformal metric, *on* N *consists of a conformal inner-product at each point* $p \in N$, *varying smoothly in an obvious sense.*

Note that this emphasizes the pointwise data of the conformal structure, unlike the usual definition of a conformal structure as an equivalence class of global Riemannian metrics. An easy topological argument shows that these notions are equivalent.

3.1.1 Flat Conformal Space

We start with oriented Lorentz space $\mathbf{L}^{n+2}$, with coordinates $x = (x^0, \ldots, x^{n+1})$, orientation

$$dx^0 \wedge \cdots \wedge dx^{n+1} > 0,$$

and inner-product

$$\langle x, y \rangle = -(x^0 y^{n+1} + x^{n+1} y^0) + \sum_i x^i y^i.$$

Throughout this section, we use the index ranges $0 \leq a, b \leq n+1$ and $1 \leq i, j \leq n$.

A non-zero vector $x \in \mathbf{L}^{n+2}$ is *null* if $\langle x, x \rangle = 0$. A null vector x is *positive* if $x^0 > 0$ or $x^{n+1} > 0$; this designation is often called a "time-orientation" for $\mathbf{L}^{n+2}$. The symmetries of Lorentz space are the linear transformations of $\mathbf{L}^{n+2}$ preserving the inner-product, the orientation, and the time-orientation, and they constitute a connected Lie group $SO^o(n+1, 1)$. We denote the space of positive null vectors by

$$Q = \{x \in \mathbf{L}^{n+2} : \langle x, x \rangle = 0, \text{ and } x^0 > 0 \text{ or } x^{n+1} > 0\},$$

which is one half of the familiar light-cone, with axis $\{x^i = x^0 - x^{n+1} = 0\}$.

We now define *flat conformal space* R to be the space of null lines in $\mathbf{L}^{n+2}$. As a manifold, R is a non-singular quadric in the projective space $\mathbf{P}(\mathbf{L}^{n+2})$, which is preserved by the natural action of the symmetry group $SO^o(n+1,1)$ of $\mathbf{L}^{n+2}$. We will describe the flat conformal structure on R below, in terms of the Maurer-Cartan form of the group. Note that the obvious map $Q \to R$, which we will write as $x \mapsto [x]$, gives a principal bundle with structure group $\mathbf{R}^*$.

In the literature, R is usually defined as $\mathbf{R}^n$ with a point added at infinity to form a topological sphere. To make this identification, note that for $x, y \in Q$, we have $\langle x, y \rangle \leq 0$, with equality if and only if $[x] = [y]$. We then claim that

$$H_y \overset{def}{=} \{x \in Q : \langle x, y \rangle = -1\}$$

is diffeomorphic to both $\mathbf{R}^n$ and $R\backslash[y]$; this is easily proved for $y = (0, \ldots, 0, 1)$, for instance, where the map $\mathbf{R}^n \to H_y$ is given by

$$(x^1, \ldots, x^n) \mapsto (1, x^1, \ldots, x^n, \tfrac{1}{2}||x||^2). \tag{3.1}$$

The classical description of the conformal structure on R is obtained by transporting the Euclidean metric on $\mathbf{R}^n$ to H_y, and noting that for $y \neq y'$ with $[y] = [y']$, this gives unequal but conformally equivalent metrics on $R\backslash[y]$. The fact that $SO^o(n+1,1)$ acts transitively on R then implies that for $[x] \neq [y]$ the conformal structures obtained on $R\backslash[x]$ and $R\backslash[y]$ are the same.

A *Lorentz frame* is a positively oriented basis $f = (e_0, \ldots, e_{n+1})$ of $\mathbf{L}^{n+2}$, in which e_0 and e_{n+1} are positive null vectors, and for which the inner-product is (in blocks of size $1, n, 1$, like most matrices in this section)

$$\langle e_a, e_b \rangle = \begin{pmatrix} 0 & 0 & -1 \\ 0 & I_n & 0 \\ -1 & 0 & 0 \end{pmatrix}.$$

We let P denote the set of all Lorentz frames. There is a standard simply transitive right-action of $SO^o(n+1,1)$ on P, by which we can identify the two spaces in a way that depends on a choice of basepoint in P; this gives P the structure of a smooth manifold. Because we have used the *right*-action, the pullback to P of any *left*-invariant 1-form on $SO^o(n+1,1)$ is independent of this choice of basepoint. These pullbacks can be intrinsically described on P as follows. We view each e_a as a map $P \to \mathbf{L}^{n+2}$, and we define 1-forms ρ, ω^i, β_j, α^i_j on P by decomposing the $\mathbf{L}^{n+2}$-valued 1-forms de_a in terms of the bases $\{e_b\}$:

$$\begin{cases} de_0 = 2e_0\rho + e_i\omega^i, \\ de_j = e_0\beta_j + e_i\alpha^i_j + e_{n+1}\omega^j, \\ de_{n+1} = e_i\beta_i - 2e_{n+1}\rho. \end{cases}$$

Equivalently,

$$d\begin{pmatrix} e_0 & e_j & e_{n+1} \end{pmatrix} = \begin{pmatrix} e_0 & e_i & e_{n+1} \end{pmatrix} \begin{pmatrix} 2\rho & \beta_j & 0 \\ \omega^i & \alpha^i_j & \beta_i \\ 0 & \omega^j & -2\rho \end{pmatrix}.$$

These forms satisfy $\alpha^i_j + \alpha^j_i = 0$ but are otherwise linearly independent, and they span the left-invariant 1-forms on $SO^o(n+1,1)$ under the preceding identification with P. Decomposing the exterior derivatives of these equations gives the Maurer-Cartan equations, expressed in matrix form as

$$d\begin{pmatrix} 2\rho & \beta_j & 0 \\ \omega^i & \alpha^i_j & \beta_i \\ 0 & \omega^j & -2\rho \end{pmatrix} + \begin{pmatrix} 2\rho & \beta_k & 0 \\ \omega^i & \alpha^i_k & \beta_i \\ 0 & \omega^k & -2\rho \end{pmatrix} \wedge \begin{pmatrix} 2\rho & \beta_j & 0 \\ \omega^k & \alpha^k_j & \beta_k \\ 0 & \omega^j & -2\rho \end{pmatrix} = 0. \quad (3.2)$$

All of the local geometry of R that is invariant under $SO^o(n+1,1)$ can be expressed in terms of these Maurer-Cartan forms. In particular, the fibers of the map $\pi_R : P \to R$ given by

$$\pi_R : (e_0, \ldots, e_{n+1}) \mapsto [e_0]$$

are the integral manifolds of the integrable Pfaffian system

$$I_R = \{\omega^1, \ldots, \omega^n\}.$$

This fibration has the structure of a principal bundle, whose structure group consists of matrices in $SO^o(n+1,1)$ of the form

$$g = \begin{pmatrix} r^2 & b_j & \frac{1}{2}r^{-2}\sum b_j^2 \\ 0 & a^i_j & r^{-2}a^i_k b_k \\ 0 & 0 & r^{-2} \end{pmatrix}, \quad (3.3)$$

where $r > 0$, $a^i_k a^j_k = \delta^{ij}$. Now, the symmetric differential form on P given by

$$q = \sum(\omega^i)^2$$

is semibasic for $\pi_R : P \to R$, and a Lie derivative computation using the structure equations (3.2) gives, for any vertical vector field $v \in \mathrm{Ker}\,(\pi_R)_*$,

$$\mathcal{L}_v q = 4(v \lrcorner \rho)q.$$

This implies that there is a unique conformal structure $[ds^2]$ on R whose representative metrics pull back under π_R^* to multiples of q. By construction, this conformal structure is invariant under the action of $SO^o(n+1,1)$, and one can verify that it gives the same structure as the classical construction described above.

In §3.1.2, we will follow Cartan in showing that associated to any conformal structure $(N, [ds^2])$ is a principal bundle $P \to N$ with 1-forms $\alpha^i_j = -\alpha^j_i$, ρ, ω^i, and β_j, satisfying structure equations like (3.2) but with generally non-zero curvature terms on the right-hand side.

Before doing this, however, we point out a few more structures in the flat model which will have useful generalizations. These correspond to Pfaffian systems

$$I_R = \{\omega^i\}, \ I_Q = \{\omega^i, \rho\}, \ I_M = \{\omega^i, \rho, \beta_j\}, \ I_{P_0} = \{\omega^i, \rho, \alpha^i_j\},$$

each of which is integrable, and in fact has a global quotient; that is, there are manifolds R, Q, M, and P_0, and surjective submersions from P to each of these, whose leaves are the integral manifolds of I_R, I_Q, I_M, and I_{P_0}, respectively:

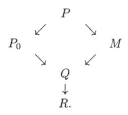

We have already seen that the leaves of the system I_R are fibers of the map $\pi_R : P \to R$. Similarly, the leaves of I_Q are fibers of the map $\pi_Q : P \to Q$ given by

$$\pi_Q : (e_0, \ldots, e_{n+1}) \mapsto e_0.$$

To understand the leaves of I_M, we let M be the set of ordered pairs (e, e') of positive null vectors satisfying $\langle e, e' \rangle = -1$. We then have a surjective submersion $\pi_M : P \to M$ defined by

$$\pi_M : (e_0, \ldots, e_{n+1}) \mapsto (e_0, e_{n+1}),$$

and the fibers of this map are the leaves of the Pfaffian system I_M. Note that the 1-form ρ and its exterior derivative are semibasic for $\pi_M : P \to M$, and this means that there is a 1-form (also called ρ) on M which pulls back to $\rho \in \Omega^1(P)$. In fact, the equation for $d\rho$ in (3.2) shows that on P,

$$\rho \wedge (d\rho)^n \neq 0,$$

so the same is true on M. Therefore, ρ defines an $SO^o(n+1, 1)$-invariant contact structure on M. The reader can verify that M has the structure of an $\mathbf{R}^*$-bundle over the space $G^{(1,1)}(\mathbf{L}^{n+2})$ parameterizing those oriented 2-planes in $\mathbf{L}^{n+2}$ on which the Lorentz metric has signature $(1, 1)$. In this context, $2\rho \in \Omega^1(M)$ can be interpreted as a connection 1-form.

Finally, to understand the leaves of I_{P_0}, we proceed as follows. Define a *conformal frame* for $(R, [ds^2])$ at a point $[x] \in R$ to be a positive basis $(v_1, \ldots, v_n)$ for $T_{[x]}R$ normalizing the conformal inner-product as

$$ds^2(v_i, v_j) = \lambda \delta_{ij},$$

for some $\lambda \in \mathbf{R}^*$ not depending on i, j. The set of conformal frames for $(R, [ds^2])$ is the total space of a principal bundle $P_0 \to R$, and there is a surjective submersion $P \to P_0$. This last is induced by the maps $\overline{e_i} : P \to TR$ associating to a Lorentz frame $f = (e_0, \ldots, e_{n+1})$ an obvious tangent vector $\overline{e_i}$ to R at $[e_0]$. The reader can verify that the fibers of the map $P \to P_0$ are the leaves of the Pfaffian system I_{P_0}.

Each of the surjective submersions $P \to R$, $P \to Q$, $P \to M$, $P \to P_0$ has the structure of a principal bundle, defined as a quotient of P by a subgroup of

$SO^o(n+1,1)$. Additionally, the spaces P, R, Q, M, and P_0 are homogeneous spaces of $SO^o(n+1,1)$, induced by the standard left-action on $\mathbf{L}^{n+2}$.

We conclude with a brief description of the geometry of $SO^0(n+1,1)$ acting on flat conformal space R. This will be useful later in understanding the space of conservation laws of conformally invariant Euler-Lagrange equations. There are four main types of motions.

- The *translations* are defined as motions of R induced by left-multiplication by matrices of the form

$$\begin{pmatrix} 1 & 0 & 0 \\ w^i & I_n & 0 \\ \frac{\|w\|^2}{2} & w^j & 1 \end{pmatrix}.$$ (3.4)

In the standard coordinates on $R\backslash\{\infty\}$ described in (3.1), this is simply translation by the vector (w^i).

- The *rotations* are defined as motions of R induced by matrices of the form

$$\begin{pmatrix} 1 & 0 & 0 \\ 0 & a^i_j & 0 \\ 0 & 0 & 1 \end{pmatrix},$$

where $(a^i_j) \in SO(n, \mathbf{R})$. In the standard coordinates, this is the usual rotation action of the matrix (a^i_j).

- The *dilations* are defined as motions of R induced by matrices of the form

$$\begin{pmatrix} r^2 & 0 & 0 \\ 0 & I & 0 \\ 0 & 0 & r^{-2} \end{pmatrix}.$$

In the standard coordinates, this is dilation about the origin by a factor of r^{-2}.

- The *inversions* are defined as motions of R induced by matrices of the form

$$\begin{pmatrix} 1 & b_j & \frac{\|b\|^2}{2} \\ 0 & I & b_i \\ 0 & 0 & 1 \end{pmatrix}.$$

Note that these are exactly conjugates of the translation matrices (3.4) by the matrix

$$J = \begin{pmatrix} 0 & 0 & 1 \\ 0 & I & 0 \\ 1 & 0 & 0 \end{pmatrix}.$$

Now, J itself is not in $SO^o(n+1,1)$, but it still acts in an obvious way on R; in standard coordinates, it gives the familiar inversion in the sphere of radius $\sqrt{2}$. So the inversions can be thought of as conjugates of translation by the standard sphere-inversion, or alternatively, as "translations with the origin fixed."

These four subgroups generate $SO^o(n+1,1)$. Although the conformal isometry group of R has more than this one component, the others do not appear in the Lie algebra, so they do not play a role in calculating conservation laws for conformally invariant Euler-Lagrange equations.

3.1.2 The Conformal Equivalence Problem

We will now apply the method of equivalence to conformal structures of dimension $n \geq 3$. This will involve some of the ideas used in the equivalence problem for definite Poincaré-Cartan forms discussed in the preceding chapter, but we we will also encounter the new concept of *prolongation*. This is the step that one takes when the usual process of absorbing and normalizing the torsion in a G-structure does not uniquely determine a pseudo-connection.

Let $(N, [ds^2])$ be an oriented conformal manifold of dimension $n \geq 3$, and let $P_0 \to N$ be the bundle of 0th-*order oriented conformal coframes* $\omega = (\omega^1, \ldots, \omega^n)$, which by definition satisfy

$$[ds^2] = \left[\sum (\omega^i)^2\right], \quad \omega^1 \wedge \cdots \wedge \omega^n > 0.$$

This is a principal bundle with structure group

$$CO(n, \mathbf{R}) = \{A \in GL^+(n, \mathbf{R}) : A\,{}^t A = \lambda I, \text{ for some } \lambda \in \mathbf{R}^*\},$$

having Lie algebra

$$
\begin{aligned}
\mathfrak{co}(n, \mathbf{R}) &= \{a \in \mathfrak{gl}(n, \mathbf{R}) : a + {}^t a = \lambda I, \text{ for some } \lambda \in \mathbf{R}\} \\
&= \{(-2r\delta^i_j + a^i_j) : a^i_j + a^j_i = 0, \ a^i_j, r \in \mathbf{R}\}.
\end{aligned}
$$

We will describe a principal bundle $P \to P_0$, called the *prolongation* of $P_0 \to N$, whose sections correspond to torsion-free pseudo-connections in $P_0 \to N$, and construct a canonical parallelism of P which defines a *Cartan connection* in $P \to N$. In case $(N, [ds^2])$ is isomorphic to an open subset of flat conformal space, this will correspond to the restriction of the Lorentz frame bundle $P \to R$ to that open subset, with parallelism given by the Maurer-Cartan forms of $SO^o(n+1,1) \cong P$.

Recall that a pseudo-connection in $P_0 \to N$ is a $\mathfrak{co}(n, \mathbf{R})$-valued 1-form

$$\varphi = (\varphi^i_j) = (-2\rho\delta^i_j + \alpha^i_j), \qquad \alpha^i_j + \alpha^j_i = 0,$$

whose restriction to each tangent space of a fiber of $P_0 \to N$ gives the canonical identification with $\mathfrak{co}(n, \mathbf{R})$ induced by the group action. As discussed previously (see §2.1), this last requirement means that φ satisfies a structure equation

$$d\omega^i = -\varphi^i_j \wedge \omega^j + \tfrac{1}{2} T^i_{jk} \omega^j \wedge \omega^k, \quad T^i_{jk} + T^i_{kj} = 0, \tag{3.5}$$

where ω^i are the components of the tautological $\mathbf{R}^n$-valued 1-form on P_0, and $\tfrac{1}{2} T^i_{jk} \omega^j \wedge \omega^k$ is the semibasic $\mathbf{R}^n$-valued torsion 2-form. We also noted previously that a pseudo-connection φ is a genuine connection if and only if it is Ad-equivariant for the action of $CO(n, \mathbf{R})$ on P_0, meaning that

$$R_g^* \varphi = Ad_{g^{-1}}(\varphi),$$

where $R_g : P_0 \to P_0$ is the right-action of $g \in CO(n, \mathbf{R})$ and $Ad_{g^{-1}}$ is the adjoint action on $\mathfrak{co}(n, \mathbf{R})$, where φ takes its values. However, completing this equivalence problem requires us to consider the more general notion of a pseudo-connection. Although the parallelism that we eventually construct is sometimes called the "conformal connection," there is no canonical way (that is, no way that is invariant under all conformal automorphisms) to associate to a conformal structure a linear connection in the usual sense.

What we seek instead is a pseudo-connection φ_j^i for which the torsion vanishes, $T_{jk}^i = 0$. We know from the fundamental lemma of Riemannian geometry, which guarantees a unique torsion-free connection in the orthonormal frame bundle of any Riemannian manifold, that whatever structure equation (3.5) we have with some initial pseudo-connection, we can alter the pseudo-connection-forms $\alpha_j^i = -\alpha_i^j$ to arrange that $T_{jk}^i = 0$. Specifically, we replace

$$\alpha_j^i \rightsquigarrow \alpha_j^i + \tfrac{1}{2}(T_{jk}^i - T_{ik}^j - T_{ij}^k)\omega^k.$$

So we can assume that $T_{jk}^i = 0$, and we have simply

$$d\omega^i = -\varphi_j^i \wedge \omega^j = -(-2\rho\delta_j^i + \alpha_j^i) \wedge \omega^j,$$

with $\alpha_j^i + \alpha_i^j = 0$. However, in contrast to Riemannian geometry, this condition on the torsion does not uniquely determine the pseudo-connection forms ρ, α_j^i. If we write down an undetermined semibasic change in pseudo-connection

$$\begin{cases} \rho \rightsquigarrow \rho + t_k \omega^k, \\ \alpha_j^i \rightsquigarrow \alpha_j^i + t_{jk}^i \omega^k, \quad t_{jk}^i + t_{ik}^j = 0, \end{cases}$$

then the condition that the new pseudo-connection be torsion-free is that

$$(2\delta_j^i t_k - t_{jk}^i)\omega^j \wedge \omega^k = 0.$$

This boils down eventually to the condition

$$t_{jk}^i = 2(\delta_k^j t_i - \delta_k^i t_j).$$

Therefore, given one torsion-free pseudo-connection φ_j^i in $P_0 \to N$, the most general is obtained by adding

$$2(\delta_j^i t_k - \delta_k^j t_i + \delta_k^i t_j)\omega^k, \tag{3.6}$$

where $t = (t_k) \in \mathbf{R}^n$ is arbitrary. This fact is needed for the next step of the equivalence method, which consists of prolonging our $CO(n, \mathbf{R})$-structure. We now digress to explain this general concept, starting with the abstract machinery underlying the preceding calculation.

We begin by amplifying the discussion of normalizing torsion in §2.1. Associated to any linear Lie algebra $\mathfrak{g} \subset \mathfrak{gl}(n, \mathbf{R})$ is an exact sequence of $\mathfrak{g}$-modules

$$0 \to \mathfrak{g}^{(1)} \to \mathfrak{g} \otimes (\mathbf{R}^n)^* \overset{\delta}{\to} \mathbf{R}^n \otimes \bigwedge\nolimits^2 (\mathbf{R}^n)^* \to H^{0,1}(\mathfrak{g}) \to 0. \qquad (3.7)$$

Here, the map δ is the restriction to the subspace

$$\mathfrak{g} \otimes (\mathbf{R}^n)^* \subset (\mathbf{R}^n \otimes (\mathbf{R}^n)^*) \otimes (\mathbf{R}^n)^*$$

of the surjective skew-symmetrization map

$$\mathbf{R}^n \otimes (\mathbf{R}^n)^* \otimes (\mathbf{R}^n)^* \to \mathbf{R}^n \otimes \bigwedge\nolimits^2 (\mathbf{R}^n)^*.$$

The space $\mathfrak{g}^{(1)}$ is the kernel of this restriction, and is called the *prolongation* of $\mathfrak{g}$; the cokernel $H^{0,1}(\mathfrak{g})$, a *Spencer cohomology group* of $\mathfrak{g}$, was encountered in §2.1. Note that $\mathfrak{g}^{(1)}$ and $H^{0,1}(\mathfrak{g})$ depend on the representation $\mathfrak{g} \hookrightarrow \mathfrak{gl}(n, \mathbf{R})$, and not just on the abstract Lie algebra $\mathfrak{g}$.

Recall from §2.1 that the intrinsic torsion of a G-structure vanishes if and only if there exist (locally) torsion-free pseudo-connections in that G-structure. This is a situation in which further canonical reduction of the structure group is not generally possible. In particular, this will always occur for G-structures with $H^{0,1}(\mathfrak{g}) = 0$.

In this situation, the torsion-free pseudo-connection is unique if and only if $\mathfrak{g}^{(1)} = 0$. For example, when $\mathfrak{g} = \mathfrak{so}(n, \mathbf{R})$, both $\mathfrak{g}^{(1)} = 0$ and $H^{0,1}(\mathfrak{g}) = 0$, which accounts for the existence and uniqueness of a torsion-free, metric-preserving connection on any Riemannian manifold. In this favorable situation, we have essentially completed the method of equivalence, because the tautological form and the unique torsion-free pseudo-connection constitute a canonical, global coframing for the total space of our G-structure. Equivalences of G-structures correspond to isomorphisms of the associated coframings, and there is a systematic procedure for determining when two parallelized manifolds are locally isomorphic.

However, one frequently works with a structure group for which $\mathfrak{g}^{(1)} \neq 0$. The observation that allows us to proceed in this case is that any pseudo-connection φ in a G-structure $P \to N$ defines a particular type of $\mathfrak{g} \oplus \mathbf{R}^n$-valued coframing

$$\varphi \oplus \omega : TP \to \mathfrak{g} \oplus \mathbf{R}^n \qquad (3.8)$$

of the total space P. Our previous discussion implies that given some torsion-free pseudo-connection φ, any change φ' lying in $\mathfrak{g}^{(1)} \subset \mathfrak{g} \otimes (\mathbf{R}^n)^*$ yields a pseudo-connection $\varphi + \varphi'$ which is also torsion-free. This means that the coframings of P as in (3.8), with φ torsion-free, are exactly the sections of a $\mathfrak{g}^{(1)}$-structure $P^{(1)} \to P$, where we regard $\mathfrak{g}^{(1)}$ as an abelian Lie group. This $P^{(1)} \to P$ is by definition the *prolongation* of the G-structure $P \to N$, and differential invariants of the former are also differential invariants of the latter.[1] The next natural step in studying $P \to N$ is therefore to start over with $P^{(1)} \to P$, by choosing a pseudo-connection, absorbing and normalizing its torsion, and so forth.

In practice, completely starting over would be wasteful. The total space $P^{(1)}$ supports tautological forms φ and ω, valued in $\mathfrak{g}$ and $\mathbf{R}^n$, respectively; and the equation $d\omega + \varphi \wedge \omega = 0$ satisfied by any particular torsion-free pseudo-connection φ on P still holds on $P^{(1)}$ with φ replaced by a tautological form. We can therefore differentiate this equation and try to extract results about the algebraic form of $d\varphi$. These results can be interpreted as statements about the intrinsic torsion of $P^{(1)} \to P$. Only then do we return to the usual normalization process. We will now illustrate this, returning to our situation in the conformal structure equivalence problem.

We have shown the existence of torsion-free pseudo-connections φ_j^i in the $CO(n, \mathbf{R})$-structure $P_0 \to N$, so the intrinsic torsion of $P_0 \to N$ vanishes.[2] We also have that such φ_j^i are unique modulo addition of a semibasic $\mathfrak{co}(n, \mathbf{R})$-valued 1-form linearly depending on an arbitrary choice of $(t_k) \in \mathbf{R}^n$. Therefore $\mathfrak{co}(n, \mathbf{R})^{(1)} \cong \mathbf{R}^n$, and the inclusion $\mathfrak{co}(n, \mathbf{R})^{(1)} \hookrightarrow \mathfrak{co}(n, \mathbf{R}) \otimes (\mathbf{R}^n)^*$ is described by (3.6). As explained above, we have an $\mathbf{R}^n$-structure $P \stackrel{\text{def}}{=} (P_0)^{(1)} \to P_0$, whose sections correspond to torsion-free pseudo-connections in $P_0 \to N$. Any choice of the latter trivializes $P \to P_0$, and then $(t_k) \in \mathbf{R}^n$ is a fiber coordinate. We now search for structure equations on P, with the goal of identifying a canonical $\mathbf{R}^n$-valued pseudo-connection form for $P \to P_0$.

The first structure equation is still

$$dw^i = -\varphi_j^i \wedge \omega^j,$$

where

$$\varphi_j^i = -2\delta_j^i \rho + \alpha_j^i, \quad \alpha_j^i + \alpha_i^j = 0,$$

and α_j^i, ρ are tautological forms on P. Differentiating this gives

$$(d\varphi_j^i + \varphi_k^i \wedge \varphi_j^k) \wedge \omega^j = 0, \tag{3.9}$$

so

$$d\varphi_j^i + \varphi_k^i \wedge \varphi_j^k \equiv 0 \quad (\text{mod } \{\omega^1, \ldots, \omega^n\}).$$

Taking the trace of this equation of matrix 2-forms shows that $d\rho \equiv 0$, so guided by the flat model (3.2), we write

$$d\rho = -\tfrac{1}{2}\beta_i \wedge \omega^i \tag{3.10}$$

---

[1] Situations with non-unique torsion-free pseudo-connections are not the only ones that call for prolongation; sometimes one finds intrinsic torsion lying in the fixed set of $H^{0,1}(\mathfrak{g})$, and essentially the same process being described here must be used. However, we will not face such a situation.

[2] In fact, what we proved is that δ is surjective for $\mathfrak{g} = \mathfrak{co}(n, \mathbf{R})$, so $H^{0,1}(\mathfrak{co}(n, \mathbf{R})) = 0$.

for some 1-forms β_i which are not uniquely determined. We will recognize these below as pseudo-connection forms in $P \to P_0$, to be uniquely determined by conditions on the torsion which we will uncover shortly. Substituting (3.10) back into (3.9), we have

$$(d\alpha^i_j + \alpha^i_k \wedge \alpha^k_j - \beta_j \wedge \omega^i + \beta_i \wedge \omega^j) \wedge \omega^j = 0,$$

and we set

$$A^i_j = d\alpha^i_j + \alpha^i_k \wedge \alpha^k_j - \beta_j \wedge \omega^i + \beta_i \wedge \omega^j.$$

Note that $A^i_j + A^j_i = 0$. We can write

$$A^i_j = \psi^i_{jk} \wedge \omega^k$$

for some 1-forms $\psi^i_{jk} = \psi^i_{kj}$, in terms of which the condition $A^i_j + A^j_i = 0$ is

$$(\psi^i_{jk} + \psi^j_{ik}) \wedge \omega^k = 0,$$

which implies

$$\psi^i_{jk} + \psi^j_{ik} \equiv 0 \quad (\mathrm{mod}\ \{\omega^1, \ldots, \omega^n\}).$$

Now computing modulo $\{\omega^1, \ldots, \omega^n\}$ as in Riemannian geometry, we have

$$\psi^i_{jk} \equiv -\psi^j_{ik} \equiv -\psi^j_{ki} \equiv \psi^k_{ji} \equiv \psi^k_{ij} \equiv -\psi^i_{kj} \equiv -\psi^i_{jk}, \tag{3.11}$$

so $\psi^i_{jk} \equiv 0$. We can now write

$$A^i_j = \psi^i_{jk} \wedge \omega^k = \tfrac{1}{2} A^i_{jkl} \omega^k \wedge \omega^l,$$

and forget about the ψ^i_{jk}, as our real interest is in $d\alpha^i_j$. We can assume that $A^i_{jkl} + A^i_{jlk} = 0$, and we necessarily have $A^i_{jkl} + A^j_{ikl} = 0$. Substituting once more into $A^i_j \wedge \omega^j = 0$, we find that

$$A^i_{jkl} + A^i_{klj} + A^i_{ljk} = 0.$$

In summary, we have

$$d\alpha^i_j + \alpha^i_k \wedge \alpha^k_j - \beta_j \wedge \omega^i + \beta_i \wedge \omega^j = \tfrac{1}{2} A^i_{jkl} \omega^k \wedge \omega^l,$$

where A^i_{jkl} has the symmetries of the Riemann curvature tensor.

In particular, we need only n new 1-forms β_i to express the derivatives of $d\rho$, $d\alpha^i_j$. The β_i are pseudo-connection forms for the prolonged $\mathfrak{co}(n, \mathbf{R})^{(1)}$-bundle $P \to P_0$, chosen to eliminate torsion in the equation for $d\rho$, while the functions A^i_{jkl} constitute the torsion in the equations for $d\alpha^i_j$. Some of this torsion will now be absorbed in the usual manner, by making a uniquely determined choice of β_i.

Notice that the equation (3.10) for $d\rho$ is preserved exactly under substitutions of the form

$$\beta_i \rightsquigarrow \beta_i + s_{ij}\omega^j, \qquad s_{ij} = s_{ji}.$$

This substitution will induce a change

$$A^i_{jkl} \rightsquigarrow A^i_{jkl} + (-\delta^i_l s_{jk} + \delta^j_l s_{ik} + \delta^i_k s_{jl} - \delta^j_k s_{il}).$$

Now, we know from the symmetries of the Riemann curvature tensor that

$$A^l_{jkl} = A^l_{kjl},$$

and on this contraction (the "Ricci" component) our substitution will induce the change

$$A^l_{jkl} \rightsquigarrow A^l_{jkl} - (n-2)s_{jk} - \delta_{jk}s_{ll}.$$

As we are assuming $n \geq 3$, there is a unique choice of s_{ij} which yields

$$A^l_{jkl} = 0.$$

It is not difficult to compute that the appropriate s_{ij} is given by

$$s_{ij} = \tfrac{1}{n-2}\left(A^l_{ijl} - \tfrac{1}{2n-2}\delta_{ij}A^l_{kkl}\right).$$

In summary,

On P, there is a unique coframing $\omega^i, \rho, \beta_j, \alpha^i_j = -\alpha^j_i$, where ω^i are the tautological forms over N, and such that the following structure equations are satisfied:

$$\begin{aligned}
d\omega^i &= (2\delta^i_j\rho - \alpha^i_j) \wedge \omega^j, \\
d\rho &= -\tfrac{1}{2}\beta_i \wedge \omega^i, \\
d\alpha^i_j &= -\alpha^i_k \wedge \alpha^k_j + \beta_j \wedge \omega^i - \beta_i \wedge \omega^j + \tfrac{1}{2}A^i_{jkl}\omega^k \wedge \omega^l, \\
&\quad \text{with } A^l_{jkl} = 0.
\end{aligned}$$

We now seek structure equations for $d\beta_j$. We start by differentiating the simplest equation in which β_j appears, which is $d\rho = -\tfrac{1}{2}\beta_j \wedge \omega^j$, and this gives

$$(d\beta_j + 2\rho \wedge \beta_j + \beta_k \wedge \alpha^k_j) \wedge \omega^j = 0.$$

We write

$$d\beta_j + 2\rho \wedge \beta_j + \beta_k \wedge \alpha^k_j = B_{jk} \wedge \omega^k \qquad (3.12)$$

for some 1-forms $B_{jk} = B_{kj}$. Because the equation for $d\rho$ did not determine β_j uniquely, we cannot expect to use it to completely determine expressions for $d\beta_j$; we need to differentiate the equations for $d\alpha^i_j$, substituting (3.12). This gives

$$(DA^i_{jkl} - B_{ik}\delta^j_l + B_{jk}\delta^i_l + B_{il}\delta^j_k - B_{jl}\delta^i_k) \wedge \omega^k \wedge \omega^l = 0.$$

Here we have defined for convenience the "covariant derivative"

$$DA^i_{jkl} = dA^i_{jkl} + 4\rho A^i_{jkl} + \alpha^i_m A^m_{jkl} - A^i_{mkl}\alpha^m_j - A^i_{jml}\alpha^m_k - A^i_{jkm}\alpha^m_l. \qquad (3.13)$$

Now we can write

$$DA^i_{jkl} - B_{ik}\delta^j_l + B_{jk}\delta^i_l + B_{il}\delta^j_k - B_{jl}\delta^i_k \equiv 0 \pmod{\{\omega^1,\dots,\omega^n\}},$$

and contracting on il gives

$$B_{jk} \equiv 0 \pmod{\{\omega^1,\dots,\omega^n\}}.$$

This allows us to write simply

$$d\beta_i + 2\rho \wedge \beta_i + \beta_j \wedge \alpha^j_i = \tfrac{1}{2}B_{ijk}\omega^j \wedge \omega^k,$$

for some functions $B_{ijk} = -B_{ikj}$. Returning to the equation

$$0 = d^2\rho = -\tfrac{1}{2}d(\beta_j \wedge \omega^j)$$

now yields the cyclic symmetry

$$B_{ijk} + B_{jki} + B_{kij} = 0.$$

We now have complete structure equations, which can be summarized in the matrix form suggested by the flat model (3.2):

$$\phi \stackrel{def}{=} \begin{pmatrix} 2\rho & \beta_j & 0 \\ \omega^i & \alpha^i_j & \beta_i \\ 0 & \omega^j & -2\rho \end{pmatrix}, \quad \Phi \stackrel{def}{=} d\phi + \phi \wedge \phi = \begin{pmatrix} 0 & B_j & 0 \\ 0 & A^i_j & B_i \\ 0 & 0 & 0 \end{pmatrix}, \quad (3.14)$$

where

$$A^i_j = \tfrac{1}{2}A^i_{jkl}\omega^k \wedge \omega^l,$$
$$A^i_{jkl} + A^j_{ikl} = A^i_{jkl} + A^i_{jlk} = 0,$$
$$A^i_{jkl} + A^i_{klj} + A^i_{ljk} = A^l_{jkl} = 0,$$
$$B_j = \tfrac{1}{2}B_{jkl}\omega^k \wedge \omega^l,$$
$$B_{jkl} + B_{jlk} = B_{jkl} + B_{klj} + B_{ljk} = 0.$$

Furthermore, the action of $\mathbf{R}^n$ on $P \to P_0$ and that of $CO(n,\mathbf{R})$ on $P_0 \to N$ may be combined, to realize $P \to N$ as a principal bundle having structure group $G \subset SO^o(n+1,1)$ consisting of matrices of the form (3.3). The matrix 1-form ϕ in (3.14) defines an $\mathfrak{so}(n+1,1)$-valued parallelism on P, under which the tangent spaces of fibers of $P \to N$ are carried to the Lie algebra $\mathfrak{g} \subset \mathfrak{so}(n+1,1)$ of G, and ϕ is equivariant with respect to the adjoint action of G on $\mathfrak{so}(n+1,1)$. The data of $(P \to N, \phi)$ is often called a *Cartan connection* modeled on $\mathfrak{g} \hookrightarrow \mathfrak{so}(n+1,1)$.

We conclude this discussion by describing some properties of the functions A^i_{jkl}, B_{jkl} on P. Differentiating the definition of Φ (3.14) yields the *Bianchi identity*

$$d\Phi = \Phi \wedge \varphi - \varphi \wedge \Phi.$$

The components of this matrix equation yield linear-algebraic consequences about the derivatives of A^i_{jkl}, B_{jkl}. First, one finds that

$$\tfrac{1}{2} D A^i_{jkl} \wedge \omega^k \wedge \omega^l = \tfrac{1}{2} B_{ikl} \omega^j \wedge \omega^k \wedge \omega^l - \tfrac{1}{2} B_{jkl} \omega^i \wedge \omega^k \wedge \omega^l. \tag{3.15}$$

Detailed information can be obtained from this equation, but note immediately the fact that

$$D A^i_{jkl} \equiv 0 \pmod{\{\omega^i\}}.$$

In particular, referring to the definition (3.13), this shows that the collection of functions (A^i_{jkl}) vary along the fibers of $P \to N$ by a linear representation of the structure group G. In other words, they correspond to a section of an associated vector bundle over N. Specifically, we can see that the expression

$$A \overset{def}{=} \tfrac{1}{4} A^i_{jkl} (\omega^i \wedge \omega^j \otimes \omega^k \wedge \omega^l) \otimes (\omega^1 \wedge \cdots \wedge \omega^n)^{-2/n}$$

on P is invariant under the group action, so A defines a section of

$$\mathrm{Sym}^2(\textstyle\bigwedge^2 T^* N) \otimes D^{-2/n},$$

where D is the *density line bundle* for the conformal structure, to be defined shortly. This section is called the *Weyl tensor* of the conformal structure.

Something different happens with B_{jkl}. Namely, the Bianchi identity for dB_{jkl} yields

$$\begin{aligned} DB_{jkl} &\overset{def}{=} dB_{jkl} + 6\rho B_{jkl} - B_{mkl}\alpha^m_j - B_{jml}\alpha^m_k - B_{jkm}\alpha^m_l \\ &\equiv -\beta_i A^i_{jkl} \pmod{\{\omega^i\}}. \end{aligned}$$

In particular, the collection (B_{jkl}) transforms by a representation of G if and only if the Weyl tensor $A = 0$. In case $n = 3$, the symmetry identities of A^i_{jkl} imply that $A = 0$ automatically; there is no Weyl tensor in 3-dimensional conformal geometry. In this case, (B_{jkl}) defines a section of the vector bundle $T^* N \otimes \bigwedge^2 T^* N$, which actually lies in a subbundle, consisting of traceless elements of the kernel of

$$T^* N \otimes \textstyle\bigwedge^2 T^* N \to \bigwedge^3 T^* N \to 0.$$

This section is called the *Cotten tensor* of the 3-dimensional conformal structure. If the Cotten tensor vanishes, then the conformal structure is locally equivalent to the flat conformal structure on the 3-sphere.

In case $n > 3$, from (3.15) one can show that the functions B_{jkl} can be expressed as linear combinations of the covariant derivatives of A^i_{jkl}. In particular, if the Weyl tensor A vanishes, then so do all of the B_{jkl}, and the conformal structure of N is locally equivalent to the flat conformal structure on the n-sphere.

3.1.3 The Conformal Laplacian

To every conformal manifold $(N^n, [ds^2])$ is canonically associated a linear differential operator Δ, called the *conformal Laplacian*. In this section, we define this operator and discuss its elementary properties. One subtlety is that Δ does not act on functions, but on sections of a certain *density line bundle*, and our first task is to define this. We will use the parallelized principal bundle $\pi : P \to N$ canonically associated to $[ds^2]$ as in the preceding discussion.

To begin, note that any n-form σ on N pulls back to P to give a closed n-form

$$\pi^* \sigma = u \, \omega^1 \wedge \cdots \wedge \omega^n \in \Omega^n(P),$$

where u is a function on P whose values on a fiber $\pi^{-1}(x)$ give the coefficient of $\sigma_x \in \bigwedge^n(T_x^* N)$ with respect to various conformal coframes at $x \in N$. Among all n-forms on P of the form $u \, \omega^1 \wedge \cdots \wedge \omega^n$, those that are locally pullbacks from N are characterized by the property of being closed. Using the structure equations, we find that this is equivalent to

$$(du + 2nu\rho) \wedge \omega^1 \wedge \cdots \wedge \omega^n = 0,$$

or

$$du \equiv -2nu\rho \pmod{\{\omega^1, \ldots, \omega^n\}}.$$

This is the infinitesimal form of the relation

$$u(p \cdot g) = r^{-2n} u(p), \tag{3.16}$$

for $p \in P$ and $g \in G$ as in (3.3). This is in turn the same as saying that the function u on P defines a section of the oriented line bundle $D \to N$ associated to the 1-dimensional representation $g \mapsto r^{2n}$ of the structure group.[3] Positive sections of D correspond to oriented volume forms on N, which in an obvious way correspond to Riemannian metrics representing the conformal class $[ds^2]$. Because so many of the PDEs studied in the conformal geometry literature describe conditions on such a metric, we should expect our study of Euler-Lagrange equations in conformal geometry to involve this density bundle.

In analogy with this, we define for any positive real number s the degree-$\frac{s}{n}$ density bundle $D^{s/n}$ associated to the 1-dimensional representation $g \mapsto r^{2s}$; the degree-1 density bundle is the preceding D. Sections are represented by functions u on P satisfying

$$u(p \cdot g) = r^{-2s} u(p), \tag{3.17}$$

or infinitesimally,

$$du \equiv -2su\rho \pmod{\{\omega^1, \ldots, \omega^n\}}. \tag{3.18}$$

---

[3]That is, D is the quotient of $P \times \mathbf{R}$ by the equivalence relation $(p, u) \sim (p \cdot g, r^{-2n}u)$ for $p \in P$, $u \in \mathbf{R}$, $g \in G$; a series of elementary exercises shows that this is naturally a line bundle over N, whose sections correspond to functions $u(p)$ satisfying (3.16).

Summarizing, we will say that any function u on P satisfying (3.18) defines a section of the degree-$\frac{s}{n}$ density bundle, and write

$$u \in \Gamma(D^{s/n}).$$

We further investigate the local behavior of $u \in \Gamma(D^{s/n})$, writing

$$du + 2su\rho = u_i\omega^i$$

for some "first covariant derivative" functions u_i. Differentiating again, and applying the Cartan lemma, we obtain

$$du_i + su\beta_i + 2(s+1)u_i\rho - u_j\alpha_i^j = u_{ij}\omega^j, \tag{3.19}$$

for some "second covariant derivatives" $u_{ij} = u_{ji}$; this is the infinitesimal form of the transformation rule

$$u_i(p \cdot g) = r^{-2(s+1)}(u_j(p)a_i^j - sb_iu(p)). \tag{3.20}$$

Note that unless $s = 0$ (so that u is actually a function on N), the vector-valued function (u_i) on P does not represent a section of any associated vector bundle.

Differentiating again, and factoring out ω^k, we obtain modulo $\{\omega^1, \ldots, \omega^n\}$

$$du_{ij} \equiv \delta_{ij}u_k\beta_k - (s+1)(u_j\beta_i + u_i\beta_j) - 2(s+2)u_{ij}\rho + u_{kj}\alpha_i^k + u_{ik}\alpha_j^k,$$

so once again, u_{ij} is not a section of any associated vector bundle. However, we can take the trace

$$du_{ii} \equiv (n - 2s - 2)\beta_k u_k - 2(s+2)\rho u_{ii} \pmod{\{\omega^1, \ldots, \omega^n\}},$$

and we see that in case $s = \frac{n-2}{2}$, the function u_{ii} on P is a section of $D^{\frac{s+2}{n}}$. To summarize,

the map $u \mapsto u_{ii}$ defines a second-order linear differential operator, called the conformal Laplacian,

$$\Delta : \Gamma(D^{\frac{n-2}{2n}}) \to \Gamma(D^{\frac{n+2}{2n}}).$$

Note that for sections $u, v \in \Gamma(D^{\frac{n-2}{2n}})$, the quantity $u\Delta v \in \Gamma(D^1)$ can be thought of as an n-form on N, and integrated. Furthermore, the reader can compute that

$$(u\Delta v - v\Delta u)\omega = d((uv_i - vu_i)\omega_{(i)}).$$

We interpret this as saying that $u\Delta v - v\Delta u$ is canonically a divergence, so that $(\cdot, \Delta\cdot)$ is a symmetric bilinear form on $\Gamma_o(D^{\frac{n-2}{2n}})$, where

$$(\cdot, \cdot) : \Gamma_o(D^{\frac{n-2}{2n}}) \times \Gamma_o(D^{\frac{n+2}{2n}}) \to \mathbf{R}$$

is given by integration on N of the product.

To clarify the meaning of Δ, we can choose a particular Riemannian metric g representing the conformal structure, and compare the second covariant derivatives of an $\frac{s}{n}$-density u taken in the conformal sense with those derivatives taken in the usual sense of Riemannian geometry. By construction of P, the pulled-back quadratic form $\pi^* g \in \mathrm{Sym}^2(T^*P)$ may be expressed as

$$\pi^* g = \lambda((\omega^1)^2 + \cdots + (\omega^n)^2)$$

for some function $\lambda > 0$ on P. Proceeding in a manner similar to the preceding, we note that

$$\mathcal{L}_v(\pi^* g) = 0$$

for any vector field v that is vertical for $P \to N$. Knowing the derivatives of ω^i quite explicitly, we can then calculate that

$$d\lambda = -4\lambda\rho + \lambda_i \omega^i \tag{3.21}$$

for some functions λ_i. Differentiating again, we find

$$d\lambda_i = -2\lambda\beta_i - 6\lambda_i\rho + \lambda_j\alpha_i^j + \lambda_{ij}\omega^j, \tag{3.22}$$

for some functions $\lambda_{ij} = \lambda_{ji}$. Now we can reduce our bundle $P \to N$ to a subbundle $P_g \subset P$, defined by

$$P_g = \{p \in P : \lambda(p) = 1, \ \lambda_1(p) = \cdots = \lambda_n(p) = 0\}.$$

Equations (3.21, 3.22) imply that P_g has structure group $SO(n, \mathbf{R}) \subset G$, and using bars to denote restrictions to P_g, we have for the pseudo-connection forms

$$\bar{\rho} = 0, \quad \bar{\beta}_i = \tfrac{1}{2}\bar{\lambda}_{ij}\bar{\omega}^j, \quad d\bar{\omega}^i = -\bar{\alpha}_j^i \wedge \bar{\omega}^j.$$

The last of these means that if we identify P_g with the usual orthonormal frame bundle of (N, g), then $\bar{\alpha}_j^i$ gives the Levi-Civita connection. The curvature is by definition

$$d\bar{\alpha}_j^i + \bar{\alpha}_k^i \wedge \bar{\alpha}_j^k = \tfrac{1}{2}R^i_{jkl}\bar{\omega}^k \wedge \bar{\omega}^l,$$

but we have an expression for the left-hand side coming from the conformal geometry; namely,

$$d\bar{\alpha}_j^i + \bar{\alpha}_k^i \wedge \bar{\alpha}_j^k = -\bar{\beta}_i \wedge \bar{\omega}^j + \bar{\beta}_j \wedge \bar{\omega}^i + \tfrac{1}{2}A^i_{jkl}\bar{\omega}^k \wedge \bar{\omega}^l.$$

Substituting $\beta_i = \tfrac{1}{2}\bar{\lambda}_{ij}\bar{\omega}^j$ and comparing these two expressions gives

$$R^i_{jkl} = \tfrac{1}{2}(\delta^i_l\bar{\lambda}_{jk} - \delta^j_l\bar{\lambda}_{ik} - \delta^i_k\bar{\lambda}_{jl} + \delta^j_k\bar{\lambda}_{il}) + A^i_{jkl}.$$

From this we find the other components of curvature

$$\mathrm{Ric}_{jl} = R^i_{jil} = \tfrac{1}{2}((2 - n)\bar{\lambda}_{jl} - \delta^j_l\bar{\lambda}_{ii}),$$
$$R = \mathrm{Ric}_{ll} = (1 - n)\bar{\lambda}_{ii}.$$

Now we will compute the conformal Laplacian of $u \in \Gamma(D^{\frac{n-2}{2n}})$, but restrict the computation to P_g. Note that the choice of g amounts to a trivialization of D and of all of its powers, so in this setting it is correct to think of u as a function. We have

$$
\begin{aligned}
d\bar{u} &= -(n-2)\bar{\rho}\bar{u} + \bar{u}_i\bar{\omega}^i \\
&= \bar{u}_i\bar{\omega}^i, \\
d\bar{u}_i &= -\tfrac{n-2}{2}\bar{u}\bar{\beta}_i - n\bar{u}_i\bar{\rho} + \bar{u}_j\bar{\alpha}_i^j + \bar{u}_{ij}\bar{\omega}^j \\
&= \bar{u}_j\bar{\alpha}_i^j + (\bar{u}_{ij} - \tfrac{n-2}{4}\bar{u}\bar{\lambda}_{ij})\bar{\omega}^j.
\end{aligned}
$$

Denoting by Δ_g the Riemannian Laplacian, we now have

$$
\begin{aligned}
\Delta_g \bar{u} &= \bar{u}_{ii} - \tfrac{n-2}{4}\bar{u}\bar{\lambda}_{ii} \\
&= \Delta\bar{u} + \tfrac{n-2}{4(n-1)}R\bar{u}.
\end{aligned}
$$

This is the more familiar expression for the conformal Laplacian, defined in terms of the Riemannian Laplacian of some representative metric. In the case of the flat model of conformal geometry, if one uses standard coordinates on $\mathbf{R}^n = R\backslash\{\infty\}$, then the Euclidean metric represents the conformal class, and we can use the ordinary Laplacian $\Delta = \sum(\frac{\partial}{\partial x^i})^2$. Its transformation properties, often stated and proved with tedious calculations, can be easily derived from the present viewpoint.

Of particular interest to us will be non-linear Poisson equations, of the form

$$
\Delta u = f(x^i, u), \tag{3.23}
$$

where we will have an interpretation of Δ as the conformal Laplacian on a conformal manifold with coordinates x^i. We will therefore want to interpret the unknown u as a section of the density bundle $D^{\frac{n-2}{2n}}$, and we will want to interpret $f(x, u)$ as a $(0^{th}$-order) bundle map

$$
f : D^{\frac{n-2}{2n}} \to D^{\frac{n+2}{2n}}.
$$

Certain obvious bundle maps f come to mind. One kind is given by multiplication by any section $\lambda \in \Gamma(D^{2/n})$; this would make (3.23) a linear equation. Another is the appropriate power map

$$
u \mapsto u^{\frac{n+2}{n-2}}.
$$

This yields a non-linear Poisson equation, and we will examine it quite closely in what follows.

We conclude this discussion with an alternate perspective on the density bundles $D^{s/n}$. First, note that for any conformal manifold $(N, [ds^2])$ with its associated parallelized bundle $P \to N$, the Pfaffian system

$$
I_Q = \{\rho, \omega^1, \dots, \omega^n\}
$$

is integrable, and its associated foliation is simple. The leaf space of this foliation is just the quotient Q of P by the action of a subgroup of its structure group, and this Q is also a fiber bundle over N, with fiber $\mathbf{R}^*$. This generalizes the space Q of positive null vectors in $\mathbf{L}^{n+2}$ which appeared in the discussion of the flat model. Now, the density bundles $D^{s/n}$ are all canonically oriented, and we claim that Q is canonically identified with the *positive* elements of $D^{s/n}$, for any s.

To see this, note that any positive $u \in D^{s/n}$, over $x \in N$, is defined as a positive function on the fiber $P_x \subset P$ satisfying (3.17). It is not hard to see that the locus $\{p \in P_x : u(p) = 1\} \subset P_x$ is a leaf of the foliation defined by I_Q. Conversely, let $L_Q \subset P$ be a leaf of the foliation defined by I_Q. Then L_Q lies completely in some fiber P_x of $P \to N$, and we can define a function u on P_x by setting $u = 1$ on L_Q, and extending to P_x by the rule (3.17). These are clearly inverse processes.

We can extend the identification as follows. Let $J^1(N, D_+^{s/n})$ be the space of 1-jets of positive sections of D; it is a contact manifold, in the usual manner. Let M be the leaf space of the simple foliation associated to the integrable Pfaffian system on P

$$I_M = \{\rho, \omega^1, \ldots, \omega^n, \beta_1, \ldots, \beta_n\}.$$

This M is also a contact manifold, with global contact form pulling back to $\rho \in \Omega^1(P)$, and it generalizes the contact manifold M mentioned in our discussion of the flat model. We claim that there is a canonical contact isomorphism between $J^1(N, D_+^{s/n})$ and M.

To see this, note that a 1-jet at $x \in N$ of a positive section of $D^{s/n}$ is specified by $n + 1$ functions $(u, u_1, \ldots, u_n)$ on the fiber P_x satisfying (3.17, 3.20). It is then not hard to see that the locus $\{p \in P_x : u(p) = 1, \ u_i(p) = 0\} \subset P_x$ is a leaf of the foliation defined by I_M. Conversely, let $L_M \subset P$ be a leaf of the foliation defined by I_M. Then L_M lies completely in some fiber P_x of $P \to N$, and we can define $n + 1$ functions $(u, u_1, \ldots, u_n)$ on P_x by setting $u = 1$ and $u_i = 0$ on L_M, and extending to P_x by the rules (3.17, 3.20). These are again inverse processes, and we leave it to the reader to investigate the correspondence between contact structures.

3.2 Conformally Invariant Poincaré-Cartan Forms

In this section, we identify the Poincaré-Cartan forms on the contact manifold M over flat conformal space R that are invariant under the action of the conformal group $SO^o(n + 1, 1)$. We then specialize to one that is neo-classical, and determine expressions for the corresponding Euler-Lagrange equation in coordinates; it turns out to be the non-linear Poisson equation with critical exponent

$$\Delta u = C u^{\frac{n+2}{n-2}}.$$

The calculation should clarify some of the more abstract constructions of the preceding section. It will also be helpful in understanding the branch of the equivalence problem in which this Poincaré-Cartan form appears, which is the topic of the next section.

We denote by P the set of Lorentz frames for $\mathbf{L}^{n+2}$, by M the set of pairs (e, e') of positive null vectors with $\langle e, e' \rangle = -1$, by Q the space of positive null vectors, and by R the flat conformal space of null lines. There are $SO^o(n+1, 1)$-equivariant maps

$$
\begin{cases}
\pi_M : P \to M, & (e_0, \dots, e_{n+1}) \mapsto (e_0, e_{n+1}), \\
\pi_Q : P \to Q, & (e_0, \dots, e_{n+1}) \mapsto e_0, \\
\pi_R : P \to R, & (e_0, \dots, e_{n+1}) \mapsto [e_0].
\end{cases}
$$

For easy reference we recall the structure equations for Lorentz frames

$$
\begin{cases}
de_0 = 2e_0\rho + e_i\omega^i, \\
de_j = e_0\beta_j + e_i\alpha^i_j + e_{n+1}\omega^j, & \alpha^i_j + \alpha^j_i = 0, \\
de_{n+1} = e_i\beta_i - 2e_{n+1}\rho;
\end{cases}
\tag{3.24}
$$

$$
\begin{cases}
d\rho + \frac{1}{2}\beta_i \wedge \omega^i = 0, \\
d\omega^i - 2\rho \wedge \omega^i + \alpha^i_j \wedge \omega^j = 0, \\
d\beta_i + 2\rho \wedge \beta_i + \beta_j \wedge \alpha^j_i = 0, \\
d\alpha^i_j + \alpha^i_k \wedge \alpha^k_j + \beta_i \wedge \omega^j - \beta_j \wedge \omega^i = 0.
\end{cases}
\tag{3.25}
$$

We noted in the previous section that M has a contact 1-form which pulls back to ρ, and this is the setting for our Poincaré-Cartan forms.

Proposition 3.1 *The $SO^o(n+1, 1)$-invariant Poincaré-Cartan forms on M, pulled back to P, are constant linear combinations of*

$$
\Pi_k \overset{def}{=} \rho \wedge \sum_{|I|=k} \beta_I \wedge \omega_{(I)},
$$

where $0 \le k \le n$. Those that are neo-classical with respect to Q are of the form

$$
\Pi = c_1\Pi_1 + c_0\Pi_0, \quad c_0, c_1 \in \mathbf{R}.
\tag{3.26}
$$

Proof. In this setting, an invariant Poincaré-Cartan form on M, pulled back to P, is an $(n+1)$-form that is a multiple of ρ, semibasic over M, invariant under the left-action of $SO^o(n+1, 1)$, invariant under the right-action of the isotropy subgroup $SO(n, \mathbf{R})$ of M, and closed. That Π must be semibasic and $SO^o(n+1, 1)$-invariant forces it to be a constant linear combination of exterior products of ρ, β_i, ω^i. It is then a consequence of the Weyl's theory of vector invariants that the further conditions of being a multiple of ρ and $SO(n, \mathbf{R})$-invariant force Π to be a linear combination of the given Π_k. It follows from the structure equations of P that $d\Pi_k = 0$, so each Π_k is in fact the pullback of a Poincaré-Cartan form. $\qquad\square$

We note that for the n-form Λ_k defined by

$$\Lambda_k \overset{def}{=} \sum_{|I|=k} \beta_I \wedge \omega_{(I)}$$

we have

$$d\Lambda_k = 2(n - 2k)\Pi_k.$$

This means that for $k \neq \frac{n}{2}$, the Poincaré-Cartan form Π_k is associated to an $SO^o(n+1,1)$-invariant functional, which in the standard coordinates discussed below is second-order. For the exceptional case $n = 2k$, there is no invariant functional corresponding to Π_k, but in the neo-classical case $k \leq 1$ with $n \geq 3$, this is not an issue.

We now focus on the neo-classical case (3.26), for which it will be convenient to rescale and study

$$\boxed{\Pi = \rho \wedge \left(\beta_i \wedge \omega_{(i)} - \frac{2C}{n-2}\omega\right),} \tag{3.27}$$

where C is a constant. This is the exterior derivative of the Lagrangian

$$\Lambda = \frac{1}{2(n-2)}\beta_i \wedge \omega_{(i)} - \frac{C}{n(n-2)}\omega,$$

and our Monge-Ampere differential system is generated by ρ and the n-form

$$\Psi = \beta_i \wedge \omega_{(i)} - \frac{2C}{n-2}\omega.$$

Proposition 3.2 *The Euler-Lagrange equation corresponding to the Poincaré-Cartan form (3.27) is locally equivalent to*

$$\Delta u = Cu^{\frac{n+2}{n-2}}. \tag{3.28}$$

The meaning of "locally equivalent" will come out in the proof. It includes an explicit and computable correspondence between integral manifolds of the Monge-Ampere system and solutions to the PDE.

We remark that the PDEs corresponding to higher Poincaré-Cartan forms Π_k, with $k > 1$, have been computed and analyzed by J. Viaclovsky in [Via00].

Proof. We begin by defining a map $\sigma : J^1(\mathbf{R}^n, \mathbf{R}) \hookrightarrow P$, which can be projected to M to give an open inclusion of contact manifolds with dense image. This map will be expressed in terms of the usual contact coordinates (x^i, z, p_i) on $J^1(\mathbf{R}^n, \mathbf{R})$, except that z is replaced by $u = e^{\lambda z}$ for some undetermined constant $\lambda \neq 0$, so that in particular,

$$dz - p_i dx^i = (\lambda u)^{-1} du - p_i dx^i.$$

When using coordinates (x^i, u, p_i), we denote our jet space by $J^1(\mathbf{R}^n, \mathbf{R}^+)$. We then pull back Ψ via σ, and consider its restriction to a transverse Legendre submanifold. With a convenient choice of λ, we will obtain a non-zero multiple of $\Delta u - C u^{\frac{n+2}{n-2}}$, implying the Proposition.

We define σ as a lift of the following map $\mathbf{R}^n \hookrightarrow P$, to be extended to $J^1(\mathbf{R}^n, \mathbf{R}^+)$ shortly:

$$\bar{e}_0(x) = \begin{pmatrix} 1 \\ x^1 \\ \vdots \\ x^n \\ \frac{\|x\|^2}{2} \end{pmatrix}, \quad \bar{e}_i(x) = \begin{pmatrix} 0 \\ \vdots \\ 1_i \\ \vdots \\ x^i \end{pmatrix}, \quad \bar{e}_{n+1}(x) = \begin{pmatrix} 0 \\ 0 \\ \vdots \\ 0 \\ 1 \end{pmatrix}. \tag{3.29}$$

It is easy to verify that this does take values in P. Also, note that the composition $\mathbf{R}^n \hookrightarrow P \to R$ gives standard (stereographic) coordinates on $R \backslash \{\infty\}$. This partly indicates the notion of "locally equivalent" used in this Proposition. We now let

$$\begin{aligned} e_0(x, u, p) &= u^{2k} \bar{e}_0(x), \\ e_i(x, u, p) &= \bar{e}_i(x) + p_i \bar{e}_0(x), \\ e_{n+1}(x, u, p) &= u^{-2k}(\bar{e}_{n+1}(x) + p_j \bar{e}_j(x) + \tfrac{\|p\|^2}{2} \bar{e}_0(x)), \end{aligned} \tag{3.30}$$

for some constant $k \neq 0$ to be determined shortly. Our use of the dependent variable u as a scaling factor for e_0 reflects the fact that we expect u to represent a section of some density line bundle. The formula for e_{n+1} is chosen just so that our map takes values in P.

Now we can compute directly

$$\begin{aligned} de_0 &= 2ku^{-1}e_0 du + u^{2k}\bar{e}_i dx^i \\ &= 2(ku^{-1}du - \tfrac{1}{2}p_i dx^i)e_0 + (u^{2k}dx^i)e_i, \end{aligned}$$

so by comparison with the expression in (3.24) we obtain some of the pulled-back Maurer-Cartan forms:

$$\begin{aligned} \sigma^* \omega^i &= u^{2k} dx^i, \\ \sigma^* \rho &= ku^{-1}du - \tfrac{1}{2}p_i dx^i. \end{aligned}$$

Similarly, we have

$$\begin{aligned} \beta_i &= -\langle e_{n+1}, de_i \rangle \\ &= -\langle u^{-2k}(\bar{e}_{n+1} + p_j \bar{e}_j + \tfrac{\|p\|^2}{2}\bar{e}_0), \\ &\qquad \bar{e}_{n+1} dx^i + p_i \bar{e}_k dx^k + \bar{e}_0 dp_i \rangle \\ &= u^{-2k}\left(dp_i - p_i p_j dx^j + \tfrac{\|p\|^2}{2}dx^i\right). \end{aligned}$$

Because we want the projection to M of $\sigma : J^1(\mathbf{R}^n, \mathbf{R}^+) \to P$ to be a contact mapping, we need $\sigma^*\rho$ to be a multiple of $dz - p_i dx^i = (\lambda u)^{-1}du - p_i dx^i$, which holds if we choose

$$k = \tfrac{1}{2\lambda}.$$

Now, λ is still undetermined, but it will shortly be chosen to simplify the expression for the restriction of Π to a transverse Legendre submanifold. Namely, we find that

$$\beta_i \wedge \omega_{(i)} = u^{\frac{n-2}{\lambda}} \left(dp_i \wedge dx_{(i)} + \tfrac{n-2}{2}||p||^2 dx \right),$$

and also

$$\omega = u^{\frac{n}{\lambda}} dx.$$

On transverse Legendre submanifolds, we have

$$du = \lambda e^{\lambda z} dz = \lambda u p_i dx^i,$$

so that

$$p_i = \frac{1}{\lambda u} \frac{\partial u}{\partial x^i}.$$

Differentiating, we obtain

$$dp_i = \frac{1}{\lambda} \left(\frac{1}{u} \frac{\partial^2 u}{\partial x^i \partial x^j} - \frac{1}{u^2} \frac{\partial u}{\partial x^i} \frac{\partial u}{\partial x^j} \right) dx^j,$$

so that on transverse Legendre submanifolds,

$$\Psi = u^{\frac{n-2}{\lambda}} \left(\frac{1}{\lambda} \frac{\Delta u}{u} + \frac{n-2-2\lambda}{2\lambda^2} \frac{||\nabla u||^2}{u^2} - \left(\frac{2C}{n-2} \right) u^{2/\lambda} \right) dx = 0.$$

We can eliminate the first-order term by choosing

$$\lambda = \tfrac{n-2}{2},$$

and then

$$\Psi = \frac{2u}{n-2} \left(\Delta u - C u^{\frac{n+2}{n-2}} \right) dx,$$

which is the desired result. $\square$

Note that $z = \lambda^{-1} \log u$ satisfies a PDE that is slightly more complicated, but equivalent under a classical transformation. Also, note that (3.28) is usually given as the Euler-Lagrange equation of the functional

$$\int \left(\tfrac{1}{2}||\nabla u||^2 + \tfrac{n-2}{2n} C u^{\frac{2n}{n-2}} \right) dx,$$

which has the advantage of being first-order, but the disadvantage of not being preserved by the full conformal group $SO^o(n+1,1)$. In contrast, our Lagrangian Λ restricts to transverse Legendre submanifolds (in the coordinates of the preceding proof) as the variationally equivalent integrand

$$\Lambda = \left(\tfrac{1}{(n-2)^2} u \Delta u - \tfrac{C}{n(n-2)} u^{\frac{2n}{n-2}} \right) dx.$$

3.3   The Conformal Branch of the Equivalence Problem

Let (M^{2n+1}, Π) be a manifold with a non-degenerate Poincaré-Cartan form; that is, $\Pi \in \Omega^{n+1}(M)$ is closed, and has a linear divisor that is unique modulo scaling and defines a contact structure. We also assume that $n \geq 3$ and that Π is neo-classical and definite. Then as discussed in §2.4 we may associate to (M, Π) a G-structure $B \to M$, where G is a subgroup of $GL(2n + 1, \mathbf{R})$ whose Lie algebra consists of matrices of the form

$$\begin{pmatrix} (n-2)r & 0 & 0 \\ 0 & -2r\delta^i_j + a^i_j & 0 \\ d_i & s_{ij} & nr\delta^j_i - a^j_i \end{pmatrix}, \tag{3.31}$$

where $a^i_j + a^j_i = 0$ and $s_{ij} = s_{ji}$, $s_{ii} = 0$. In this section, we show how to uniquely characterize in terms of the invariants of the G-structure those (M, Π) which are locally equivalent to the Poincaré-Cartan form for the equation

$$\Delta u = C u^{\frac{n+2}{n-2}}, \qquad C \neq 0, \tag{3.32}$$

on flat conformal space. The result may be loosely summarized as follows.

> The vanishing of the primary invariants T^{ijk}, U^{ij}, S^i_j is equivalent to the existence of a foliation $B \to N$ over a conformal manifold $(N, [ds^2])$, for which $[ds^2]$ pulls back to the invariant $[\sum(\omega^i)^2]$. In this case, under open conditions on further invariants, three successive reductions of $B \to M$ yield a subbundle which is naturally identified with the conformal bundle over N. The Poincaré-Cartan form can then be identified with that associated to a non-linear Poisson equation. In case a further invariant is constant, this equation is equivalent to (3.32).

We find these conditions by continuing to apply the equivalence method begun in §2.4, pursuing the case in which all of the non-constant torsion vanishes. One corollary of the discussion is a characterization of Poincaré-Cartan forms locally equivalent to those for general non-linear Poisson equations of the form

$$\Delta u = f(x, u), \qquad x \in N, \tag{3.33}$$

on an n-dimensional conformal manifold $(N, [ds^2])$; here and in the following, Δ is the conformal Laplacian. The condition that (3.33) be non-linear can be characterized in terms of the geometric invariants associated to (M, Π), as can the condition that $(N, [ds^2])$ be conformally flat. The characterization of (3.32) will imply that this equation has maximal symmetry group among non-linear Euler-Lagrange equations satisfying certain geometric conditions on the torsion. We will not actually prove the characterization result for general Poisson equations (3.33), but we will use these equations (in the conformally flat case, with $\Delta = \sum \left(\frac{\partial}{\partial x^i} \right)^2$) as an example at each stage of the following calculations.

We first recall the structure equations of the G-structure $B \to M$, associated to a neo-classical, definite Poincaré-Cartan form

$$\Pi = -\theta \wedge (\pi_i \wedge \omega_{(i)}).$$

There is a pseudo-connection

$$\varphi = \begin{pmatrix} (n-2)\rho & 0 & 0 \\ 0 & -2\rho\delta_j^i + \alpha_j^i & 0 \\ \delta_i & \sigma_{ij} & n\rho\delta_i^j - \alpha_i^j \end{pmatrix}, \quad \text{with} \quad \begin{cases} \alpha_j^i + \alpha_i^j = 0, \\ \sigma_{ij} = \sigma_{ji}, \ \sigma_{ii} = 0, \end{cases}$$

(3.34)

having torsion

$$d\begin{pmatrix} \theta \\ \omega^i \\ \pi_i \end{pmatrix} + \varphi \wedge \begin{pmatrix} \theta \\ \omega^j \\ \pi_j \end{pmatrix} = \begin{pmatrix} -\pi_i \wedge \omega^i \\ -(S_j^i \omega^j + U^{ij}\pi_j) \wedge \theta + T^{ijk}\pi_j \wedge \omega^k \\ 0 \end{pmatrix}, \quad (3.35)$$

where enough torsion has been absorbed so that

$$T^{ijk} = T^{jik} = T^{kji}, \ T^{iik} = 0; \ U^{ij} = U^{ji}; \ S_j^i = S_j^i, \ S_i^i = 0. \quad (3.36)$$

We also recall the structure equation (2.48)

$$(n-2)d\rho = -\delta_i \wedge \omega^i - S_j^i \pi_i \wedge \omega^j + \left(\tfrac{n-2}{2n}\right) U^{ij}\sigma_{ij} \wedge \theta - t^i \pi_i \wedge \theta. \quad (3.37)$$

The equations (3.34, 3.35, 3.36, 3.37) uniquely determine the forms ρ, α_j^i, and we are still free to alter our pseudo-connection by

$$\begin{cases} \delta_i \rightsquigarrow \delta_i + b_i\theta + t_{ij}\omega^j, \text{ with } t_{ij} = t_{ji} \text{ and } t_{ii} = 0, \\ \sigma_{ij} \rightsquigarrow \sigma_{ij} + t_{ij}\theta + t_{ijk}\omega^k, \text{ with } t_{ijk} = t_{jik} = t_{kji} \text{ and } t_{iik} = 0, \end{cases} \quad (3.38)$$

requiring also

$$2nb_i + (n-2)U^{jk}t_{ijk} = 0.$$

We set up our example (3.33) by taking coordinates (x^i, u, q_i) on $M = J^1(\mathbf{R}^n, \mathbf{R})$, with contact form

$$\tilde{\theta} \overset{def}{=} du - q_i dx^i.$$

Then transverse Legendre submanifolds which are also integral manifolds of

$$\tilde{\Psi} \overset{def}{=} -dq_i \wedge dx_{(i)} + f(x, u)dx$$

correspond locally to solutions of (3.33). One can verify that the form

$$\tilde{\Pi} \overset{def}{=} \tilde{\theta} \wedge \tilde{\Psi}$$

is closed, so in particular our Poisson equation is an Euler-Lagrange equation. We find a particular 1-adapted coframing of $J^1(\mathbf{R}^n, \mathbf{R})$ as in Lemma 2.3 by writing

$$\tilde{\Pi} = -\tilde{\theta} \wedge \left((dq_i - \tfrac{f}{n}dx^i) \wedge dx_{(i)}\right),$$

and then setting

$$\begin{pmatrix} \tilde{\theta} \\ \tilde{\omega}^i \\ \tilde{\pi}_i \end{pmatrix} = \begin{pmatrix} du - q_i dx^i \\ dx^i \\ dq_i - \frac{f}{n} dx^i \end{pmatrix}.$$

It turns out that this coframing is actually a section of $B \to J^1(\mathbf{R}^n, \mathbf{R})$, as one discovers by setting

$$\tilde{\rho} = 0, \quad \tilde{\alpha}^i_j = 0, \quad \tilde{\delta}_i = -\tfrac{1}{n} f_u \tilde{\omega}^i,$$

and noting that the structure equations (3.34, 3.35) hold (with some complicated choice of $\tilde{\sigma}_{ij}$ which we will not need). In fact, (3.35) holds with torsion coefficients S^i_j, U^{ij}, T^{ijk} all vanishing, and we will see the significance of this presently.

In the general setting, we seek conditions under which the quadratic form on B

$$q \overset{\text{def}}{=} \sum (\omega^i)^2$$

can be regarded as defining a conformal structure on some quotient of B. For the appropriate quotient to exist, at least locally, the necessary and sufficient condition is that the Pfaffian system $I = \{\omega^1, \ldots, \omega^n\}$ be integrable; it is easily seen from the structure equations (and we noted in §2.4) that this is equivalent to the condition

$$U^{ij} = 0.$$

We assume this in what follows, and for convenience assume further that the foliation of B by leaves of I is simple; that is, there is a smooth manifold N and a surjective submersion $B \to N$ whose fibers are the leaves of I. Coordinates on N may be thought of as "preferred independent variables" for the contact-equivalence class of our Euler-Lagrange PDE, as indicated in §2.4.

We can now compute the Lie derivative of q under a vector field v which is vertical for $B \to N$, satisfying $v \lrcorner \omega^i = 0$; using the hypothesis $U^{ij} = 0$ and the structure equations, we find

$$\mathcal{L}_v q = 2 \left(T^{ijk}(v \lrcorner \pi_j)\omega^i \omega^k + S^i_j(v \lrcorner \theta)\omega^i \omega^j \right) + 4(v \lrcorner \rho)q.$$

It follows that if $T^{ijk} = 0$ and $S^i_j = 0$, then there is a quadratic form on N which pulls back to a non-zero multiple of q on B. A short calculation shows that the converse as true as well, so we have the following.

Proposition 3.3 *The conditions $U^{ij} = T^{ijk} = S^i_j = 0$ are necessary and sufficient for there to exist (locally) a conformal manifold $(N, [ds^2])$ and a map $B \to N$ such that the pullback to B of $[ds^2]$ is equal to $[q] = [\sum (\omega^i)^2]$.*

From now on, we assume $U^{ij} = S^i_j = T^{ijk} = 0$.

From the discussion of the conformal equivalence problem in §3.1.2, we know that associated to $(N, [ds^2])$ is the second-order conformal frame bundle $P \to N$

with global coframing $\bar{\omega}^i$, $\bar{\rho}$, $\bar{\alpha}^i_j$, $\bar{\beta}_i$ satisfying structure equations

$$\begin{cases} d\bar{\omega}^i - 2\bar{\rho} \wedge \bar{\omega}^i + \bar{\alpha}^i_j \wedge \bar{\omega}^j = 0, \\ d\bar{\rho} + \frac{1}{2}\bar{\beta}_i \wedge \bar{\omega}^i = 0, \\ d\bar{\alpha}^i_j + \bar{\alpha}^i_k \wedge \bar{\alpha}^k_j + \bar{\beta}_i \wedge \bar{\omega}^j - \bar{\beta}_j \wedge \bar{\omega}^i = \frac{1}{2}\bar{A}^i_{jkl}\bar{\omega}^k \wedge \bar{\omega}^l, \\ d\bar{\beta}_i + 2\bar{\rho} \wedge \bar{\beta}_i + \bar{\beta}_j \wedge \bar{\alpha}^j_i = \frac{1}{2}\bar{B}_{ijk}\bar{\omega}^j \wedge \bar{\omega}^k. \end{cases} \qquad (3.39)$$

Our goal is to directly relate the principal bundle $B \to M$ associated to the Poincaré-Cartan form Π on M to the principal bundle $P \to N$ associated to the induced conformal geometry on N. We shall eventually find that under some further conditions stated below, the main one of which reflects the non-linearity of the Euler-Lagrange system associated to Π, there is a canonical reduction $B_3 \to M$ of the G-structure $B \to M$ such that locally $B_3 \cong P$ as parallelized manifolds.[4] Because the canonical coframings on B_3 and P determine the bundle structure of each, we will then have shown that the subbundle $B_3 \to N$ of $B \to N$ can be locally identified with the bundle $P \to N$ associated to the conformal structure $(N, [ds^2])$.

In the special case of our Poisson equation, we have $\tilde{\omega}^i = dx^i$ as part of a section of $B \to M$, so we can already see that our quotient space $N \cong \mathbf{R}^n$ is conformally flat. This reflects the fact that the differential operator Δ in (3.33) is the conformal Laplacian for flat conformal space.

We return to the general case, and make the simplifying observation that under our hypotheses,

$$0 = d^2(\omega^1 \wedge \cdots \wedge \omega^n) = -\frac{2n}{n-2} t^i \pi_i \wedge \theta \wedge \omega^1 \wedge \cdots \wedge \omega^n,$$

so that $t^i = 0$ in the equation (3.37) for $d\rho$. We now have on B the equations

$$d\omega^i = 2\rho \wedge \omega^i - \alpha^i_j \wedge \omega^j, \qquad (3.40)$$

$$d\rho = -\frac{1}{n-2}\delta_i \wedge \omega^i. \qquad (3.41)$$

With the goal of making our structure equations on B resemble the conformal structure equations (3.39), we define

$$\beta_i \stackrel{def}{=} \left(\frac{2}{n-2}\right)\delta_i.$$

The equations for $d\omega^i$ and $d\rho$ are now formally identical to those for $d\bar{\omega}^i$ and $d\bar{\rho}$, and computing exactly as in the conformal equivalence problem, we find that

$$d\alpha^i_j + \alpha^i_k \wedge \alpha^k_j + \beta_i \wedge \omega^j - \beta_j \wedge \omega^i = \frac{1}{2}A^i_{jkl}\omega^k \wedge \omega^l,$$

for some functions A^i_{jkl} on B having the symmetries of the Riemann curvature tensor.

---

[4] As in the characterization in §2.5 of prescribed mean curvature systems, we will denote by B_1, B_2, etc., reductions of the bundle $B \to M$ associated to Π, and these are unrelated to the bundles of the same names used in the construction of B.

Of course, we want A^i_{jkl} to correspond to the Weyl tensor $\bar{A}^i_{jkl}$ of $(N, [ds^2])$, so we would like to alter our pseudo-connection forms (3.34) in a way that will give

$$A^l_{jkl} = 0.$$

Again, reasoning exactly as we did in the conformal equivalence problem, we know that there are uniquely determined functions $t_{ij} = t_{ji}$ such that replacing

$$\beta_i \rightsquigarrow \beta_i + t_{ij}\omega^j$$

accomplishes this goal. However, these may have $t_{ii} \neq 0$, meaning that we cannot make the compensating change in σ_{ij} (see (3.38)) without introducing torsion in the equation for $d\pi_i$. We proceed anyway, and now have structure equations

$$d\begin{pmatrix} \theta \\ \omega^i \\ \pi_i \end{pmatrix} + \begin{pmatrix} (n-2)\rho & 0 & 0 \\ 0 & \alpha^i_j - 2\rho\delta^i_j & 0 \\ \left(\frac{n-2}{2}\right)\beta_i & \sigma_{ij} & n\rho\delta^j_i - \alpha^j_i \end{pmatrix} \wedge \begin{pmatrix} \theta \\ \omega^j \\ \pi_j \end{pmatrix} = \begin{pmatrix} -\pi_i \wedge \omega^i \\ 0 \\ A\omega^i \wedge \theta \end{pmatrix}$$
(3.42)

where $A = \frac{n-2}{2n}t_{ii}$ is a component of the original A^i_{jkl}, analogous to scalar curvature in the Riemannian setting. Also, we have

$$d\rho = -\tfrac{1}{2}\beta_i \wedge \omega^i,$$
(3.43)

$$d\alpha^i_j + \alpha^i_k \wedge \alpha^k_j + \beta_i \wedge \omega^j - \beta_j \wedge \omega^i = \tfrac{1}{2}A^i_{jkl}\omega^k \wedge \omega^l,$$
(3.44)

with $A^l_{jkl} = 0$. These uniquely determine the pseudo-connection forms ρ, α^i_j, β_i, and leave σ_{ij} determined only up to addition of terms of the form $t_{ijk}\omega^k$, with t_{ijk} totally symmetric and trace-free.

Now that β_i is uniquely determined, we can once again mimic calculations from the conformal equivalence problem, deducing from (3.40, 3.43, 3.44) that

$$d\beta_i + 2\rho \wedge \beta_i + \beta_j \wedge \alpha^j_i = \tfrac{1}{2}B_{ijk}\omega^j \wedge \omega^k,$$
(3.45)

with $B_{ijk} + B_{ikj} = 0$, $B_{ijk} + B_{jki} + B_{kij} = 0$.

In the case of our non-linear Poisson equation (3.33), a calculation shows that the modification of $\tilde{\beta}_i = \frac{2}{n-2}\tilde{\delta}_i = 0$ is not necessary, and that with everything defined as before, we have not only (3.42), but also (3.43, 3.44, 3.45) with $A^i_{jkl} = B_{ijk} = 0$. This gives us another way of seeing that the conformal structure associated to (3.33) is flat. What will be important for us, however, is the fact that along this section of $B \to J^1(\mathbf{R}^n, \mathbf{R})$, the torsion function A is

$$\tilde{A} = \tfrac{1}{n}f_u(x, u).$$

This comes out of the calculations alluded to above.

We now begin to reduce $B \to M$, as promised. To get information about the derivative of the torsion coefficient A without knowing anything about $d\sigma_{ij}$, we consider

$$
\begin{aligned}
0 &= d^2(\pi_i \wedge \omega_{(i)}) \\
&= d((n-2)\rho \wedge \pi_i \wedge \omega_{(i)} + \tfrac{n-2}{2}\theta \wedge \beta_i \wedge \omega_{(i)} + A\theta \wedge \omega) \\
&= (dA + 4\rho A) \wedge \theta \wedge \omega.
\end{aligned}
$$

This describes the variation of the function A along the fibers of $B \to Q$, where we recall that Q is the leaf space of the integrable Pfaffian system $J_{\Pi} = \{\theta, \omega^i\}$. In particular, we can write

$$
dA + 4\rho A = A_0\theta + A_i\omega^i, \tag{3.46}
$$

for some functions A_0, A_i on B. We see that on each fiber of $B \to Q$, either A vanishes identically or A never vanishes, and we assume that the latter holds throughout B. This is motivated by the case of the Poisson equation (3.33), for which $\tilde{A} = \frac{1}{n}f_u$ (so we are assuming in particular that the zero-order term $f(x, u)$ depends on u). Because the sign of A is fixed, we assume $A > 0$ in what follows. The case $A < 0$ is similar, but the case $A = 0$ is quite different.

For the first reduction of $B \to M$, we define

$$
B_1 = \{b \in B : A(b) = \tfrac{1}{4}\} \subset B.
$$

From equation (3.46) with the assumption $A > 0$ everywhere, it is clear that $B_1 \to M$ is a principal subbundle of B, whose structure group's Lie algebra consists of matrices (3.31) with $r = 0$. Furthermore, restricted to B_1 there is a relation

$$
\rho = A_0\theta + A_i\omega^i. \tag{3.47}
$$

In the case of a Poisson equation (3.33), our section $(\tilde{\theta}, \tilde{\omega}^i, \tilde{\pi}_i)$ of $B \to J^1(\mathbf{R}^n, \mathbf{R})$ is generally not a section of $B_1 \subset B$, because we have along this section that $\tilde{A} = \frac{f_u}{n}$. However, (3.46) guides us in finding a section of B_1. Namely, we define a function $r(x, u, q) > 0$ on M by

$$
r^4 = 4\tilde{A} = \tfrac{4}{n}f_u, \tag{3.48}
$$

and then one can verify that for the coframing

$$
\begin{pmatrix} \hat{\theta} \\ \hat{\omega}^i \\ \hat{\pi}_i \end{pmatrix} \stackrel{\text{def}}{=} \begin{pmatrix} r^{2-n} & 0 & 0 \\ 0 & r^2\delta^i_j & 0 \\ 0 & 0 & r^{-n}\delta^j_i \end{pmatrix} \begin{pmatrix} \tilde{\theta} \\ \tilde{\omega}^j \\ \tilde{\pi}_j \end{pmatrix},
$$

one has the structure equation (3.42), with

$$
\hat{\rho} = r^{-1}dr = \tfrac{1}{4}f_u^{-1}df_u, \quad \hat{\beta}_i = \hat{\alpha}^i_j = 0, \quad \hat{A} = \tfrac{1}{4}.
$$

Again, we won't have any need for $\hat{\sigma}_{ij}$. Observe that along this section of B_1, $\hat{\rho} = \frac{1}{4}f_u^{-1}df_u = \hat{A}_0\hat{\theta} + \hat{A}_i\hat{\omega}^i$, so that

$$
\hat{A}_0 = \tfrac{1}{4}r^{n-2}f_u^{-1}f_{uu}, \quad \hat{A}_i = \tfrac{1}{4}r^{-2}f_u^{-1}(f_{ux^i} + f_{uu}q_i), \tag{3.49}
$$

with r given by (3.48).

Returning to the general situation on B_1, we differentiate (3.47) and find

$$(dA_0 - (n-2)\rho A_0) \wedge \theta + (dA_i + 2\rho A_i - A_j \alpha_i^j + \tfrac{1}{2}\beta_i - A_0 \pi_i) \wedge \omega^i = 0,$$

and the Cartan lemma then gives

$$dA_0 - (n-2)\rho A_0 = A_{00}\theta + A_{0i}\omega^i, \qquad (3.50)$$

$$dA_i + 2\rho A_i - A_j \alpha_i^j + \tfrac{1}{2}\beta_i - A_0 \pi_i = A_{i0}\theta + A_{ij}\omega^j, \qquad (3.51)$$

with $A_{0i} = A_{i0}$ and $A_{ij} = A_{ji}$.

We interpret (3.50) as saying that if A_0 vanishes at one point of a fiber of $B_1 \to Q$, then it vanishes everywhere on that fiber. We make the assumption that $A_0 \neq 0$; the other extreme case, where $A_0 = 0$ everywhere, gives a different branch of the equivalence problem. Note that in the case of a Poisson equation (3.33), the condition $A_0 \neq 0$ implies by (3.49) that the equation is everywhere non-linear. This justifies our decision to pursue, among the many branches of the equivalence problem within the larger conformal branch, the case $A > 0$, $A_0 \neq 0$. This justification was our main reason to carry along the example of the Poisson equation, and we will not mention it again. General calculations involving it become rather messy at this stage, but how to continue should be clear from the preceding.

Returning to the general setting, our second reduction uses (3.51), which tells us that the locus

$$B_2 = \{b \in B_1 : A_i(b) = 0\} \subset B_1$$

is a principal subbundle of $B_1 \to M$, whose structure group's Lie algebra consists of matrices (3.31) with $r = d_i = 0$. Furthermore, restricted to B_2 there are relations

$$\beta_i = 2(A_{i0}\theta + A_{ij}\omega^j + A_0\pi_i),$$

and also

$$\rho = A_0\theta.$$

With the A_i out of the way we differentiate once more, and applying the Cartan lemma find that on B_2, modulo $\{\theta, \omega^i, \pi_i\}$,

$$dA_{00} \equiv 0, \qquad (3.52)$$

$$dA_{0i} \equiv A_{0j}\alpha_i^j, \qquad (3.53)$$

$$dA_{ij} \equiv A_{kj}\alpha_i^k + A_{ik}\alpha_j^k + A_0\sigma_{ij}. \qquad (3.54)$$

We interpret (3.52) as saying that A_{00} descends to a well-defined function on M. We interpret (3.53) as saying that the vector-valued function (A_{0i}) represents a section a vector bundle associated to $B_2 \to M$. We interpret (3.54) as saying that if $A_0 = 0$, then the matrix (A_{ij}) represents a section of a vector bundle associated to $B_2 \to M$. However, we have already made the assumption that $A_0 \neq 0$ everywhere. In some examples of interest, most notably for the equation $\Delta u = C u^{\frac{n+2}{n-2}}$, the section (A_{0i}) vanishes; for a general non-linear Poisson equation, this vanishing loosely corresponds to the non-linearity being translation-invariant on flat conformal space. We will not need to make any assumptions about this quantity.

This allows us to make a third reduction. With $A_0 \neq 0$, (3.54) tells us that the locus where the trace-free part of A_{ij} vanishes,

$$B_3 = \{b \in B_2 : A^0_{ij}(b) \overset{def}{=} A_{ij}(b) - \tfrac{1}{n}\delta_{ij}A_{kk}(b) = 0\},$$

is a subbundle $B_3 \to M$ of $B_2 \to M$. In terms of (3.31), the Lie algebra of the structure group of B_3 is defined by $r = d_i = s_{ij} = 0$.

Let us summarize what we have done. Starting from the structure equations (3.34, 3.35, 3.36, 3.37) on $B \to M$ for a definite, neo-classical Poincaré-Cartan form with $n \geq 3$, we specialized to the case where the torsion satisfies

$$U^{ij} = S^i_j = T^{ijk} = 0.$$

In this case, we found that the leaf space N of the Pfaffian system $\{\omega^1, \ldots, \omega^n\}$ has a conformal structure pulling back to $[\sum(\omega^i)^2]$. We replaced each pseudo-connection form δ_i by its multiple β_i, and guided by computations from conformal geometry, we determined the torsion in the equation for $d\alpha^i_j$, which resembled a Riemann curvature tensor. This torsion's analog of scalar curvature provided our fundamental invariant A, which had first "covariant derivatives" A_0, A_i, and second "covariant derivatives" A_{00}, $A_{i0} = A_{0i}$, $A_{ij} = A_{ji}$. With the assumptions

$$A \neq 0, \quad A_0 \neq 0,$$

we were able to make successive reductions by passing to the loci where

$$A = \tfrac{1}{4}, \quad A_i = 0, \quad A_{ij} = \tfrac{1}{n}\delta_{ij}A_{kk}.$$

This leaves us on a bundle $B_3 \to M$ with a coframing $\omega^i, \rho, \beta_i, \alpha^i_j$, satisfying structure equations exactly like those on the conformal bundle $P \to N$ associated with $(N, [ds^2])$. From here, a standard result shows that there is a local diffeomorphism $B_3 \to P$ under which the two coframings correspond. In particular, the invariants A^i_{jkl} and B_{jkl} remaining in the bundle B_3 equal the invariants named similarly in the conformal structure, so we can tell for example if the conformal structure associated to our Poincaré-Cartan form is flat.

We now write the restricted Poincaré-Cartan form,

$$
\begin{aligned}
\Pi &= -\theta \wedge (\pi_i \wedge \omega_{(i)}) \\
&= -\tfrac{1}{A_0}\rho \wedge \tfrac{1}{A_0}\left(\tfrac{1}{2}\beta_i - \tfrac{1}{n}A_{kk}\omega^i\right) \wedge \omega_{(i)} \\
&= -\tfrac{1}{2A_0^2}\rho \wedge (\beta_i \wedge \omega_{(i)} - 2A_{kk}\omega).
\end{aligned}
$$

We can see from previous equations that A_{kk} is constant on fibers of $B_3 \to M$. Therefore, it makes sense to say that A_{kk} is or is not constant on B_3. If it is constant, and if the conformal structure on N is flat (that is, $A^i_{jkl} = 0$ if

$n \geq 4$, or $B_{jkl} = 0$ if $n = 3$), then our Poincaré-Cartan form is equivalent to that associated to the non-linear Poisson equation

$$\Delta u = C u^{\frac{n+2}{n-2}},$$

where $C = (n-2)A_{kk}$. This completes the characterization of Poincaré-Cartan forms equivalent to that of this equation. Our next goal is to determine the conservation laws associated to this Poincaré-Cartan form.

3.4   Conservation Laws for $\Delta u = C u^{\frac{n+2}{n-2}}$

In this section, we will determine the classical conservation laws for the conformally invariant non-linear Poisson equation

$$\Delta u = C u^{\frac{n+2}{n-2}}. \tag{3.55}$$

Recall that from $\Lambda_0 = \omega$ and $\Lambda_1 = \beta_i \wedge \omega_{(i)}$ we constructed the functional

$$\Lambda = \tfrac{1}{2(n-2)}\Lambda_1 - \tfrac{C}{n(n-2)}\Lambda_0$$

having the Poincare-Cartan form

$$\Pi = d\Lambda = \rho \wedge \left(\beta_i \wedge \omega_{(i)} - \tfrac{2C}{n-2}\omega\right),$$

and that under a certain embedding $\sigma : J^1(\mathbf{R}^n, \mathbf{R}^+) \hookrightarrow P$, the Euler-Lagrange system of Π restricted to a transverse Legendre submanifold is generated by

$$\Psi = \left(\Delta u - C u^{\frac{n+2}{n-2}}\right) dx,$$

for coordinates on $J^1(\mathbf{R}^n, \mathbf{R}^+)$ described in the proof of Proposition 3.2. We also proved that the composition of $\sigma : J^1(\mathbf{R}^n, \mathbf{R}^+) \hookrightarrow P$ with the projection $P \to M$ gives an open contact embedding of $J^1(\mathbf{R}^n, \mathbf{R}^+)$ as a dense subset of M. Our invariant forms on P pull back via σ to give the following forms on $J^1(\mathbf{R}^n, \mathbf{R}^+)$, expressed in terms of the canonical coordinates (x^i, u, p_i):

$$\begin{cases} \rho = \tfrac{1}{n-2}u^{-1}du - \tfrac{1}{2}p_i dx^i, \\ \omega^i = u^{\frac{2}{n-2}} dx^i, \\ \beta_i = u^{-\frac{2}{n-2}}\left(dp_i - p_i p_j dx^j + \tfrac{\|p\|^2}{2}dx^i\right), \\ \alpha^i_j = p_j dx^i - p_i dx^j. \end{cases} \tag{3.56}$$

To describe the conservation laws, we first calculate for symmetry vector fields $V \in \mathfrak{g}_\Pi$ the expression

$$\varphi_V = V \lrcorner \Lambda \in \Omega^{n-1}(P)$$

at points of $J^1(\mathbf{R}^n, \mathbf{R}^+) \subset P$, and then restrict this $(n-1)$-form to that submanifold, where it will be a conserved integrand for the equation.

3.4.1 The Lie Algebra of Infinitesimal Symmetries

We know that the Poincaré-Cartan forms

$$\Pi_k = \rho \wedge \left(\sum_{|I|=k} \beta_I \wedge \omega_{(I)} \right)$$

on P are invariant under the simple, transitive left-action of the conformal group $SO^o(n+1,1)$. The infinitesimal generators of this action are the vector fields on P corresponding under the identification $P \cong SO^o(n+1,1)$ to *right-invariant* vector fields. Our first task is to determine the right-invariant vector fields in terms of the basis

$$\left\{ \frac{\partial}{\partial \rho}, \frac{\partial}{\partial \omega^i}, \frac{\partial}{\partial \beta_i}, \frac{\partial}{\partial \alpha^i_j} \right\}$$

of left-invariant vector fields dual to the basis of left-invariant 1-forms used previously; this is because the Maurer-Cartan equation in our setup allows us to compute only in terms of left-invariant objects.

For an unknown vector field

$$V = g\frac{\partial}{\partial \rho} + V^i \frac{\partial}{\partial \omega^i} + V_i \frac{\partial}{\partial \beta_i} + V^i_j \frac{\partial}{\partial \alpha^i_j} \qquad (V^i_j + V^j_i = 0) \qquad (3.57)$$

to be right-invariant is equivalent to the conditions

$$\mathcal{L}_V \rho = \mathcal{L}_V \omega^i = \mathcal{L}_V \beta_i = \mathcal{L}_V \alpha^i_j = 0; \qquad (3.58)$$

that is, the flow of V should preserve all left-invariant 1-forms. We will solve the system (3.58) of first-order differential equations for V along the submanifold $J^1(\mathbf{R}^n, \mathbf{R}^+) \subset P$. Such V are not generally tangent to $J^1(\mathbf{R}^n, \mathbf{R}^+)$, but the calculation of conservation laws as $V \lrcorner \Lambda$ is still valid, as $J^1(\mathbf{R}^n, \mathbf{R}^+)$ is being used only as a slice of the foliation $P \to M$. The solution will give the coefficient functions g, V^i, V_i of (3.57) in terms of the coordinates (x^i, u, p_i) of $J^1(\mathbf{R}^n, \mathbf{R}^+)$. We will not need the coefficients V^i_j, because they do not appear in $\varphi_V = V \lrcorner \Lambda$; in fact, we compute $g = V \lrcorner \rho$ only because it simplifies the rest of the solution.

First, we use the equation $\mathcal{L}_V \rho = 0$, which gives

$$\begin{aligned} 0 &= d(V \lrcorner \rho) + V \lrcorner d\rho \\ &= dg - \tfrac{1}{2}(V_i \omega^i - V^i \beta_i). \end{aligned}$$

We have the formulae (3.56) for the restrictions of ω^i and β_i to $J^1(\mathbf{R}^n, \mathbf{R}^+)$, by which the last condition becomes

$$dg = \frac{1}{2} \left(V_i u^{-\frac{2}{n-2}} dx^i - V^i u^{-\frac{2}{n-2}} (dp_i - p_i p_j dx^j + \tfrac{\|p\|^2}{2} dx^i) \right).$$

This suggests that we replace the unknowns V_i, V^i in our PDE system (3.58) with

$$v^i \stackrel{\text{def}}{=} \tfrac{1}{2} V^i u^{-\frac{2}{n-2}}, \qquad v_i \stackrel{\text{def}}{=} \tfrac{1}{2} V_i u^{\frac{2}{n-2}} - \tfrac{1}{2} V^j u^{-\frac{2}{n-2}} (-p_j p_i + \delta_{ij} \tfrac{\|p\|^2}{2}).$$

Then we have the result

$$\frac{\partial g}{\partial x^i} = v_i, \quad \frac{\partial g}{\partial p_i} = -v^i, \quad \frac{\partial g}{\partial u} = 0. \tag{3.59}$$

In particular, we now need to determine only the function g.

For this, we use the equation $\mathcal{L}_V \omega^i = 0$, which gives

$$
\begin{aligned}
0 &= d(V \lrcorner \omega^i) + V \lrcorner d\omega^i \\
&= dV^i - 2\rho V^i + \alpha_j^i V^j + 2g\omega^i - V_j^i \omega^j.
\end{aligned}
$$

When we restrict to $J^1(\mathbf{R}^n, \mathbf{R}^+)$ using (3.56) and use our new dependent variables v_i, v^i, this gives

$$dv^i = (p_i dx^j - p_j dx^i)v^j - (p_j dx^j)v^i - g\,dx^i + \tfrac{1}{2}V_j^i dx^j. \tag{3.60}$$

This says in particular that $v^i(x, u, p)$ is a function of the variables x^i alone, so along with (3.59) we find that

$$g(x, u, p) = f(x) + f^i(x)p_i,$$

for some functions $f(x)$, $f^i(x)$. Substituting this back into (3.60), we have

$$df^i = (p_i f^j - p_j f^i - \tfrac{1}{2}V_j^i + \delta_j^i f)dx^j.$$

This is a PDE system

$$\frac{\partial f^i}{\partial x^j} = p_i f^j - p_j f^i - \tfrac{1}{2}V_j^i + \delta_j^i f$$

for the unknowns $f^i(x)$, and it can be solved in the following elementary way. We first let

$$h_j^i = p_i f^j - p_j f^i - \tfrac{1}{2}V_j^i = -h_i^j$$

so that our equation is

$$\frac{\partial f^i}{\partial x^j} = h_j^i + \delta_j^i f. \tag{3.61}$$

Differentiating this with respect to x^k and equating mixed partials implies that the expression

$$\frac{\partial h_k^i}{\partial x^j} - \delta_j^i \frac{\partial f}{\partial x^k} + \delta_j^k \frac{\partial f}{\partial x^i} \tag{3.62}$$

is symmetric in j, k. It is also clearly skew-symmetric in i, k, and therefore equals zero (as in (3.11)). Now we can equate mixed partials of h_k^i to obtain

$$\delta_k^i \frac{\partial^2 f}{\partial x^j x^l} - \delta_k^j \frac{\partial^2 f}{\partial x^i x^l} = \delta_l^i \frac{\partial^2 f}{\partial x^j x^k} - \delta_l^j \frac{\partial^2 f}{\partial x^i x^k}.$$

With the standing assumption $n \geq 3$, this implies that all of these second partial derivatives of f are zero, and we can finally write

$$f(x) = r + \tfrac{1}{2}b_k x^k,$$

for some constants r, b_k. The reasons for our labeling of these and the following constants of integration will be indicated below. Because the expressions (3.62) vanish, we can integrate to obtain

$$h^i_j = -\tfrac{1}{2}a^i_j + \tfrac{1}{2}b_j x^i - \tfrac{1}{2}b_i x^j$$

for some constants $a^i_j = -a^j_i$, and then integrate (3.61) to find

$$f^i(x) = -\tfrac{1}{2}w^i + (\delta^i_j r - \tfrac{1}{2}a^i_j)x^j - \tfrac{1}{4}b_i||x||^2 + \tfrac{1}{2}\langle b, x\rangle x^i,$$

where we have written $\langle b, x\rangle = \sum b_k x^k$ and $||x||^2 = \sum(x^k)^2$. We summarize the discussion in the following.

Proposition 3.4 *The coefficients of the vector fields (3.57) on P preserving the left-invariant 1-forms ρ, ω^i along $J^1(\mathbf{R}^n, \mathbf{R}^+)$ are of the form*

$$
\begin{aligned}
g &= r + \tfrac{1}{2}\langle b, x\rangle(1 + \langle p, x\rangle) + \left(-\tfrac{1}{2}w^i + \left(\delta^i_j r - \tfrac{1}{2}a^i_j\right)x^j - \tfrac{1}{4}b_i||x||^2\right)p_i, \\
v^i &\overset{def}{=} \tfrac{1}{2}V^i u^{-\frac{2}{n-2}} = -\frac{\partial g}{\partial p_i}, \\
v_i &\overset{def}{=} \tfrac{1}{2}V_i u^{\frac{2}{n-2}} - \tfrac{1}{2}V^j u^{-\frac{2}{n-2}}\left(-p_j p_i + \delta_{ij}\frac{||p||^2}{2}\right) = \frac{\partial g}{\partial x^i},
\end{aligned}
$$

where r, b_i, w^i, $a^i_j = -a^j_i$ are constants.

It is easy to verify that such g, V^i, V_i uniquely determine $V^i_j = -V^j_i$ such that the vector field (3.57) preserves β_i and α^i_j as well, but we will not need this fact. Note that the number of constants in the Proposition equals the dimension of the Lie algebra $\mathfrak{so}(n+1, 1)$, as expected.

The reader may be aware that one should not have to solve differential equations to determine right-invariant vector fields in terms of left-invariant vector fields. In fact, an algebraic calculation will suffice, which in this case would consist of writing an arbitrary Lie algebra element

$$
g_L = \begin{pmatrix} 2r & b_j & 0 \\ w^i & a^i_j & b_i \\ 0 & w^j & -2r \end{pmatrix}
$$

interpreted as a left-invariant vector field, and conjugating by $\sigma(x, u, p) \in P$ regarded as a matrix with columns $e_0(x, u, p)$, $e_j(x, u, p)$, $e_{n+1}(x, u, p)$ given by (3.29, 3.30). The resulting $\mathfrak{so}(n+1, 1)$-valued function on $J^1(\mathbf{R}^n, \mathbf{R}^+)$ then has entries which are the coefficients of a right-invariant vector field V. The calculation is tedious, but of course the vector fields so obtained are as in Proposition 3.4.

3.4.2 Calculation of Conservation Laws

We can now use the formulae for the infinitesimal symmetries derived above to calculate the conservation laws for Π, which are $(n-1)$-forms on $J^1(\mathbf{R}^n, \mathbf{R}^+)$

that are closed when restricted to integral submanifolds of the Euler-Lagrange system.

The Noether prescription is particularly simple in this case, because the equations

$$\mathcal{L}_V \Lambda = 0 \qquad \text{and} \qquad d\Lambda = \Pi$$

mean that there are no compensating terms, and we can take for the conserved integrand just

$$\varphi_V = V \lrcorner \Lambda.$$

This is straightforward in principle, but there are some delicate issues of signs and constants. We find that for V as in Proposition 3.4,

$$V \lrcorner \Lambda_0 = V^i \omega_{(i)},$$

and restricting to $J^1(\mathbf{R}^n, \mathbf{R}^+)$, using v^i instead of V^i, we obtain

$$(V \lrcorner \Lambda_0)|_{J^1(\mathbf{R}^n, \mathbf{R}^+)} = 2u^{\frac{2n}{n-2}} v^i dx_{(i)}.$$

The analogous computation for $V \lrcorner \Lambda_1$ is a little more complicated and gives

$$(V \lrcorner \Lambda_1)_{J^1(\mathbf{R}^n, \mathbf{R}^+)} = 2u^2(-v^j dp_i \wedge dx_{(ij)} + (v_i + \tfrac{n-2}{2} v^i ||p||^2) dx_{(i)}).$$

On a transverse Legendre submanifold S of $J^1(\mathbf{R}^n, \mathbf{R}^+)$, we can use the condition $\rho = 0$ from (3.56) to write

$$p_i = \tfrac{2}{n-2} u^{-1} \tfrac{\partial u}{\partial x^i}, \tag{3.63}$$

and if we compute dp_i and $||p||^2$ for such a submanifold, then we can substitute and obtain

$$(V \lrcorner \Lambda_1)|_S = \tfrac{4}{n-2} \left(-u u_{x^i x^j} v^j + u u_{x^j x^j} v^i + u_{x^i} u_{x^j} v^j + \tfrac{n-2}{2} u^2 v_i \right) dx_{(i)}.$$

We summarize with the following.

Proposition 3.5 *The restriction of $V \lrcorner \Lambda$ to the 1-jet graph of $u(x^1, \ldots, x^n)$ equals*

$$\boxed{\varphi_V = \left(\tfrac{2}{n-2} \left(u u_{x^j x^j} v^i - u u_{x^i x^j} v^j + u_{x^i} u_{x^j} v^j \right) - \tfrac{2C}{n} u^{\frac{2n}{n-2}} v^i + u^2 v_i \right) dx_{(i)}.}$$

We now have a representative for each of the classical conservation laws corresponding to a conformal symmetry of our equation

$$\Delta u = C u^{\frac{n+2}{n-2}}. \tag{3.64}$$

We say "representative" because a conservation law is actually an equivalence class of $(n-1)$-forms. In fact, our φ_V is not the $(n-1)$-form classically taken to represent the conservation law corresponding to V; our φ_V involves second derivatives of the unknown $u(x)$, while the classical expressions are all first-order. We can find the first-order expressions by adding to φ_V a suitable exact $(n-1)$-form, obtaining

$$
\begin{aligned}
\varphi_g \overset{def}{=} {}& \varphi_V + \tfrac{2}{n-2} d(u u_{x^i} v^j \, dx_{(ij)}) \\
= {}& \Big(\tfrac{4}{n-2} u_{x^i} u_{x^j} v^j - \Big(\tfrac{2}{n-2} ||\nabla u||^2 + \tfrac{2C}{n} u^{\frac{2n}{n-2}} \Big) v^i \\
& + u^2 v_i + \tfrac{2}{n-2} u(u_{x^i} v^j_{x^j} - u_{x^j} v^i_{x^j}) \Big) \, dx_{(i)}.
\end{aligned}
$$

This turns out to give the classical expressions for the conservation laws associated to our equation (3.64), up to multiplicative constants. It could have been obtained more directly using the methods of §1.3. For this, one would work on the usual $J^1(\mathbf{R}^n, \mathbf{R})$, with standard coordinates (x^i, u, q_i) in which the contact structure is generated by

$$
\theta = du - q_i \, dx^i,
$$

and then consider the Monge-Ampere system generated by θ and

$$
\Psi = -dq_i \wedge dx_{(i)} + C u^{\frac{n+2}{n-2}} \, dx.
$$

A little experimenting yields a Lagrangian density

$$
L \, dx = \Big(\tfrac{||q||^2}{2} + \tfrac{n-2}{2n} C u^{\frac{2n}{n-2}} \Big) \, dx,
$$

so the functional

$$
\Lambda = L \, dy + \theta \wedge L_{q_i} dy_{(i)}
$$

induces the Poincaré-Cartan form

$$
\Pi = \theta \wedge \Psi = d\Lambda.
$$

One can then determine the Lie algebra of the symmetry group of Π by solving an elementary PDE system, with a result closely resembling that of Proposition 3.4. Applying the Noether prescription to these vector fields and this Λ yields $(n-1)$-forms which restrict to transverse Legendre submanifolds to give φ_g above.

Returning to our original situation, we now compute φ_g explicitly for various choices of g as in Proposition 3.4. These choices of g correspond to subgroups of the conformal group.

Translation: $g = w^i p_i$.

In this case, we have $v^i = -w^i$, $v_i = 0$, so we find on a transverse Legendre submanifold of $J^1(\mathbf{R}^n, \mathbf{R}^+)$ that

$$
\varphi_g = \Big(\tfrac{2}{n-2} ||\nabla u||^2 w^i - \tfrac{4}{n-2} u_{x^i} u_{x^j} w^j + \tfrac{2C}{n} u^{\frac{2n}{n-2}} w^i \Big) \, dx_{(i)}.
$$

The typical use of a conservation law involves its integration along the smooth $(n-1)$-dimensional boundary of a region $\Omega \subset \mathbf{R}^n$. To make more sense of the preceding expression, we take such a region to have unit normal ν and area element $d\sigma$ (with respect to the Euclidean metric), and using the fact that $q^i dx_{(i)}|_{\partial \Omega} = \langle q, \nu \rangle d\sigma$ for a vector $q = q^i \frac{\partial}{\partial x^i}$, we have

$$\varphi_g|_{\partial\Omega} = \left\langle \frac{2}{n-2}||\nabla u||^2 w - \frac{4}{n-2}\langle \nabla u, w\rangle \nabla u + \frac{2C}{n}u^{\frac{2n}{n-2}}w, \nu \right\rangle d\sigma.$$

Here, we have let $w = w^i \frac{\partial}{\partial x^i}$ be the translation vector field induced on flat conformal space $R = \mathbf{R}^n \cup \{\infty\}$ by the right-invariant vector field on P which gives this conservation law.

Rotation: $g = a^i_j p_i x^j$, $a^i_j + a^j_i = 0$.

In this case, we have $v^i = -a^i_j x^j$, $v_i = a^j_i p_j$. On a transverse Legendre submanifold of $J^1(\mathbf{R}^n, \mathbf{R}^+)$, we have from (3.63) that $p_i = \frac{2}{n-2}u^{-1}u_{x^i}$, and we find that

$$\varphi_g = \left(\left(\frac{2}{n-2}||\nabla u||^2 + \frac{2C}{n}u^{\frac{2n}{n-2}} \right) a^i_j x^j - \frac{4}{n-2}u_{x^i}u_{x^k}a^k_j x^j + \frac{2}{n-2}uu_{x^j}a^j_i \right) dx_{(i)}.$$

In this formula, the last term represents a trivial conservation law; that is, $d(uu_{x^j}a^j_i dx_{(i)}) = 0$ on any transverse Legendre submanifold, so it will be ignored below. Restricting as in the preceding case to the smooth boundary of $\Omega \subset \mathbf{R}^n$ with unit normal ν and area element $d\sigma$, this is

$$\varphi_g|_{\partial\Omega} = \left\langle \left(\frac{2}{n-2}||\nabla u||^2 + \frac{2C}{n}u^{\frac{2n}{n-2}} \right) a - \frac{4}{n-2}\langle \nabla u, a\rangle \nabla u, \nu \right\rangle d\sigma.$$

Here, we have let $a = a^i_j x^j \frac{\partial}{\partial x^i}$ be the rotation vector field induced on flat conformal space R by the right-invariant vector field on P which gives this conservation law.

Dilation: $g = 1 + x^i p_i$.

This generating function gives the right-invariant vector field whose value at the identity is the Lie algebra element (in blocks of size $1, n, 1$)

$$\begin{pmatrix} 2 & 0 & 0 \\ 0 & 0 & 0 \\ 0 & 0 & -2 \end{pmatrix},$$

which generates a 1-parameter group of dilations about the origin in flat conformal space R. In this case, we have $v^i = -x^i$, $v_i = p_i$, and on a transverse Legendre submanifold with $p_i = \frac{2}{n-2}u^{-1}u_{x^i}$, we find that

$$\varphi_g = \left(\left(\frac{2}{n-2}||\nabla u||^2 + \frac{2C}{n}u^{\frac{2n}{n-2}} \right) x^i - \frac{4}{n-2}u_{x^i}u_{x^j}x^j - 2uu_{x^i} \right) dx_{(i)}. \tag{3.65}$$

For this conservation law, it is instructive to take for $\Omega \subset \mathbf{R}^n$ the open ball of radius $r > 0$ centered at the origin, and then

$$\varphi_g|_{\partial\Omega} = \left(r \left(\tfrac{2}{n-2}||\nabla u||^2 + \tfrac{2C}{n} u^{\frac{2n}{n-2}} - \tfrac{4}{n-2}\langle \nabla u, \nu \rangle^2 \right) - 2u\langle \nabla u, \nu \rangle \right) d\sigma. \quad (3.66)$$

A simple consequence of this conservation law is the following uniqueness theorem.[5]

Theorem 3.1 (Pohožaev) *If $u(x) \in C^2(\bar{\Omega})$ is a solution to $\Delta u = C u^{\frac{n+2}{n-2}}$ in the ball Ω of radius r, with $u \geq 0$ in Ω and $u = 0$ on $\partial\Omega$, then $u = 0$.*

Proof. We will first use the conservation law to show that $\nabla u = 0$ everywhere on $\partial\Omega$. If we decompose $\nabla u = u_\tau + u_\nu \nu$ into tangential and normal components along $\partial\Omega$, so that in particular $u_\tau = 0$ by hypothesis, then the conserved integrand (3.66) is

$$\varphi_g|_{\partial\Omega} = -\tfrac{2r}{n-2} u_\nu^2 d\sigma,$$

so the conservation law $\int_{\partial\Omega} \varphi_g = 0$ implies that $u_\nu = 0$ on $\partial\Omega$.

Now with $\nabla u = 0$ on $\partial\Omega$, we can compute

$$\begin{aligned} 0 &= \int_{\partial\Omega} *du \\ &= \int_\Omega d*du \\ &= \int_\Omega \Delta u \, dx. \end{aligned}$$

But it is clear from the PDE that Δu cannot change sign, so it must vanish identically, and this implies that $u = 0$ throughout Ω. $\qquad\square$

In fact, looking at the expression (3.65) for φ_g for a more general region, it is not hard to see that the same proof applies whenever $\Omega \subset \mathbf{R}^n$ is bounded and star-shaped.

Inversion: $g = -\tfrac{1}{2} p_j b_j ||x||^2 + b_j x^j (1 + p_i x^i)$.

This is the generating function for the vector field in $R = \mathbf{R}^n \cup \{\infty\}$ which is the conjugate of a translation vector field by inversion in an origin-centered sphere.

In this case, we have $v^i = \tfrac{1}{2} b_i ||x||^2 - b_j x^j x^i$, $v_i = b_i x^j p_j - b_j x^i p_j + b_j x^j p_i + b_i$, and on a transverse Legendre submanifold, we find after some tedious calculation that

$$\begin{aligned} \varphi_g = \Big[& \left((\tfrac{2}{n-2}||\nabla u||^2 + \tfrac{2C}{n} u^{\frac{2n}{n-2}}) \delta_{ij} - \tfrac{4}{n-2} u_{x^i} u_{x^j} \right) (b_k x^k x^j - \tfrac{1}{2} b_j ||x||^2) \\ & -2u b_j x^j u_{x^i} + u^2 b_i \Big] \, dx_{(i)}. \end{aligned}$$

---

[5] See [Poh65], where a non-existence theorem is proved for a more general class of equations, for which dilation gives an integral identity instead of a conservation law.

Again taking $\Omega \subset \mathbf{R}^n$ to be the open ball of radius $r > 0$ centered at the origin, we have

$$(n-2)\varphi_g|_{\partial\Omega} = \left\langle (r^2(-4\langle\nabla u, \nu\rangle^2 + ||\nabla u||^2 + \tfrac{C}{n}u^{\frac{2n}{n-2}}) + u^2)b \right.$$
$$\left. + 2(r^2\langle b, \nabla u\rangle - (n-2)ru\langle b, \nu\rangle)\nabla u, \nu \right\rangle d\sigma,$$

where $b = b_i\frac{\partial}{\partial x^i}$ is the vector field whose conjugate by a sphere-inversion is the vector field generating the conservation law.

3.5   Conservation Laws for Wave Equations

In this section, we will consider *non-linear wave equations*

$$\Box z = f(z), \tag{3.67}$$

which are hyperbolic analogs of the non-linear Poisson equations considered previously. Here, we are working in Minkowski space $\mathbf{L}^{n+1}$ with coordinates $(t, y^1, \ldots, y^n)$, and the wave operator is

$$\Box = -\left(\frac{\partial}{\partial t}\right)^2 + \sum\left(\frac{\partial}{\partial y^i}\right)^2.$$

It is in this hyperbolic case that conservation laws have been most effectively used. Everything developed previously in this chapter for the Laplace operator and Poisson equations on Riemannian manifolds has an analog for the wave operator and wave equations on *Lorentzian* manifolds, which by definition carry a metric of signature $(n, 1)$. Indeed, even the coordinate formulae for conservation laws that we derived in the preceding section are easily altered by a sign change to give corresponding conservation laws for the wave equation. Our goal in this section is to see how certain analytic conclusions can be drawn from these conservation laws.

Before doing this, we will illustrate the usefulness of understanding the wave operator geometrically, by presenting a result of Christodoulou asserting that the Cauchy problem for the non-linear hyperbolic equation

$$\Box z = z^{\frac{n+3}{n-1}} \tag{3.68}$$

has solutions for all time, given sufficiently small initial data.[6] The proof exploits conformal invariance of the equation in an interesting way, and this is what we want to explain. Note that (3.68) is the hyperbolic analog of the maximally symmetric non-linear Poisson equation $\Delta z = z^{\frac{n+2}{n-2}}$ considered previously; the change in exponent reflects the fact that the number of independent variables is

---

[6]See [Chr86]; what is proved there is somewhat more general.

now $n+1$, instead of n. This equation will be of special interest in our discussion of conservation laws, as well.

The idea for proving the long-time existence result is to map Minkowski space $\mathbf{L}^{n+1}$, which is the domain for the unknown z in (3.68), to a *bounded* domain, in such a way that the equation (3.68) corresponds to an equation for which short-time existence of the Cauchy problem is already known. With sufficiently small initial data, the "short-time" will cover the bounded domain, and back on $\mathbf{L}^{n+1}$ we will have a global solution.

The domain to which we will map $\mathbf{L}^{n+1}$ is actually part of a *conformal compactification* of Minkowski space, analogous to the conformal compactification of Euclidean space constructed in §3.1.1. This compactification is diffeomorphic to a product $S^1 \times S^n$, and topologically may be thought of as the result of adding a point at spatial-infinity for each time, and a time-at-infinity for each spatial point. Formally, one can begin with a vector space with inner-product of signature $(n+1, 2)$, and consider the projectivized null-cone; it is a smooth, real quadric hypersurface in $\mathbf{P}^{n+2}$, which in certain homogeneous coordinates is given by

$$\xi_1^2 + \xi_2^2 = \eta_1^2 + \cdots + \eta_n^2,$$

evidently diffeomorphic to $S^1 \times S^n$. The $(n+1, 2)$-inner-product induces a Lorentz metric on this hypersurface, well-defined up to scaling, and its conformal isometry group has identity component $SO^\circ(n+1, 2)$, which we will revisit in considering conservation laws. What is important for us now is that among the representative Lorentz metrics for this conformal structure one finds

$$g = -dT^2 + dS^2,$$

where T is a coordinate on S^1, and dS^2 is the standard metric on S^n. In certain spherical coordinates ("usual" spherical coordinates applied to $\mathbf{R}^n$, after stereographic projection) this may be written

$$g = -dT^2 + dR^2 + (\sin^2 R) \, dZ^2,$$

where $R \in [0, \pi]$, and dZ^2 is the standard metric on the unit $(n-1)$-sphere.

Now we will conformally embed Minkowski space $\mathbf{L}^{n+1}$ as a bounded domain in the finite part $\mathbf{R} \times \mathbf{R}^n$ of $S^1 \times S^n$, the latter having coordinates (T, R, Z). The map $\varphi(t, r, z) = (T, R, Z)$ is given by

$$\begin{pmatrix} T \\ R \\ Z \end{pmatrix} = \begin{pmatrix} \arctan(t+r) + \arctan(t-r) \\ \arctan(t+r) \quad \arctan(t-r) \\ z \end{pmatrix},$$

and one can easily check that

$$\varphi^*(-dT^2 + dR^2 + (\sin^2 R) \, dZ^2) = \Omega^2(-dt^2 + dr^2 + r^2 dz^2),$$

where dZ^2 and dz^2 are both the standard metric on the unit $(n-1)$-sphere; the conformal factor is

$$\Omega = 2(1 + (t+r)^2)^{-\frac{1}{2}}(1 + (t-r)^2)^{-\frac{1}{2}},$$

and the right-hand side is a multiple of the flat Minkowski metric. The image of φ is the "diamond"

$$\mathcal{D} = \{(T, R, Z) : R - \pi < T < \pi - R, \ R \geq 0\}.$$

Note that the initial hyperplane $\{t = 0\}$ corresponds to $\{T = 0\}$, and that with fixed (R, Z), as $T \to \pi - R$, $t \to \infty$. Consequently, the long-time Cauchy problem for the invariant wave equation (3.68) corresponds to a short-time Cauchy problem on the bounded domain $\mathcal{D}$ for some other equation.

We can see what this other equation is without carrying out tedious calculations by considering the *conformally invariant wave operator*, an analog of the conformal Laplacian discussed in §3.1.3. This is a differential operator

$$\Box_c : \Gamma(D^{\frac{n-1}{2(n+1)}}) \to \Gamma(D^{\frac{n+3}{2(n+1)}})$$

between certain density line bundles over a manifold with Lorentz metric. With a choice of Lorentz metric g representing the conformal class, $u \in \Gamma(D^{\frac{n-1}{2(n+1)}})$ is represented a function u_g, and the density $\Box_c u$ is represented by the function

$$(\Box_c u)_g = \Box_g(u_g) + \tfrac{n-1}{4n} R_g u_g,$$

where R_g is the scalar curvature and $\Box_g$ is the wave operator associated to g. We interpret our wave equation (3.68) as a condition on a density represented in the flat (Minkowski) metric g_0 by the function u_0, and the equation transformed by the map φ introduced above should express the same condition represented in the new metric g. The representative functions are related by

$$u_g = \Omega^{-\frac{n-1}{2}} u_0, \qquad (\Box_c u)_g = \Omega^{-\frac{n+3}{2}} (\Box_c u)_0,$$

so the condition (3.68) becomes

$$\begin{aligned}
\Box_g(u_g) + \tfrac{n-1}{4n} R_g u_g &= (\Box_c u)_g \\
&= \Omega^{-\frac{n+3}{2}} (\Box_c u)_0 \\
&= \Omega^{-\frac{n+3}{2}} u_0^{\frac{n+3}{n-1}} \\
&= u_g^{\frac{n+3}{n-1}}.
\end{aligned}$$

The scalar curvature is just that of the round metric on the n-sphere, $R_g = n(n-1)$, so letting $u = u_0$, $U = u_g$, the equation (3.68) is transformed into

$$\Box_g U + \tfrac{(n-1)^2}{4} U = U^{\frac{n+3}{n-1}}. \tag{3.69}$$

Finally, suppose given compactly supported initial data $u(0, x) = u_0(x)$ and $u_t(0, x) = u_1(x)$ for (3.68). These correspond to initial data $U_0(X)$ and $U_1(X)$ for (3.69), supported in the ball of radius π. The standard result on local existence implies that the latter Cauchy problem can be solved for all X, in some time interval $T \in [0, T_0]$, with a lower-bound for T_0 determined by the size of the initial data. Therefore, with sufficiently small initial data, we can arrange $T_0 \geq \pi$, and translated back to the original coordinates, this corresponds to a global solution of (3.68).

We now turn to more general wave equations (3.67), where conservation laws have been most effectively used.[7] Equation (3.67) is the Euler-Lagrange equation for the action functional

$$\int_{\mathbf{R}} \int_{\mathbf{R}^n} \left(\tfrac{1}{2}(-z_t^2 + ||\nabla z||^2) + F(z) \right) dy\, dt,$$

where $F'(z) = f(z)$, and the gradient ∇z is with respect to the "space" variables $y^1, \ldots, y^n$.

Rather than redevelop the machinery of conformal geometry in the Lorentz case, we work in the classical setting, on $J^1(\mathbf{L}^{n+1}, \mathbf{R})$ with coordinates t, y^i, z, p_a (as usual, $1 \leq i \leq n$, $0 \leq a \leq n$), contact form $\theta = dz - p_0 dt - p_i dy^i$, Lorentz inner-product $ds^2 = -dt^2 + \sum (dy^i)^2$, and Lagrangian

$$L(t, y, z, p) = \tfrac{1}{2}(-p_0^2 + \sum |p_i|^2) + F(z).$$

A normalized representative functional is then

$$\Lambda = L\, dt \wedge dy + \theta \wedge (-p_0 dy - p_i dt \wedge dy_{(i)}), \tag{3.70}$$

satisfying

$$d\Lambda = \Pi = \theta \wedge (dp_0 \wedge dy + dp_i \wedge dt \wedge dy_{(i)} + f(z) dt \wedge dy).$$

This is the example discussed at the end of §1.3. As mentioned there, the invariance of the equation under time-translation gives an important conservation law, and its uses will be our first topic below. In fact, there are conservation laws associated to space-translations and Lorentz rotations, the latter generated by ordinary spatial rotations $b_j^i y^j \frac{\partial}{\partial y^j}$ ($b_j^i + b_i^j = 0$) and *Lorentz boosts* $y^i \frac{\partial}{\partial t} + t \frac{\partial}{\partial y^i}$; however, these seem to have been used less widely in the analysis of (3.67).

Especially interesting is the case of (3.68), which is preserved under a certain action of the *conformal Lorentz group* $SO^0(n+1, 2)$ on $J^1(\mathbf{L}^{n+1}, \mathbf{R})$. In particular, there are extensions to $J^1(\mathbf{L}^{n+1}, \mathbf{R})$ of the dilation and inversion vector fields on $\mathbf{L}^{n+1}$, and these give rise to more conservation laws. We will consider these after discussing uses of the time-translation conservation law for the more general wave equations (3.67).

3.5.1 Energy Density

The time-translation vector field $\frac{\partial}{\partial t}$ on $\mathbf{L}^{n+1}$ lifts to $J^1(\mathbf{L}^{n+1}, \mathbf{R})$ to a symmetry of Λ having the same expression, $V = \frac{\partial}{\partial t}$. The Noether prescription gives

$$\varphi_t = V \lrcorner \Lambda = \left(\tfrac{1}{2}(p_0^2 + \sum p_i^2) + F(z) \right) dy + p_0 p_i dt \wedge dy_{(i)},$$

---

[7] This material and much more may be found in [Str89].

as calculated in §1.3. The coefficient of dy here is the *energy density*

$$e \stackrel{\text{def}}{=} \tfrac{1}{2}(p_0^2 + |p_i|^2) + F(z),$$

and it appears whenever we integrate φ_t along a constant-time level surface $\mathbf{R}_t^n = \{t\} \times \mathbf{R}^n$. The energy function

$$E(t) \stackrel{\text{def}}{=} \int_{\mathbf{R}_t^n} e\, dy \geq 0,$$

is constant by virtue of (3.67), assuming sufficient decay of z and its derivatives in the space variables for the integral to make sense.

A more substantial application involves a region $\Omega \subset \mathbf{L}^{n+1}$ of the form

$$\Omega = \bigcup_{t \in (t_0, t_1)} \{\|y\| < r_0 - (t - t_0)\},$$

a union of open balls in space, with initial radius r_0 decreasing with speed 1. The boundary $\partial \Omega$ is $T - B + K$, where

- $B = \{t_0\} \times \{\|y\| \leq r_0\}$ is the initial disc,

- $T = \{t_1\} \times \{\|y\| \leq r_0 - (t_1 - t_0)\}$ is the final disc, and

- $K = \cup_{t \in [t_0, t_1]}\{\|y\| = r_0 - (t - t_0)\}$ is part of a null cone.

The conservation of φ_t on $\partial \Omega$ reads

$$0 = \int_{\partial \Omega} \varphi_t = \int_{T-B} e\, dy + \int_K \varphi_t. \tag{3.71}$$

The term $\int_K \varphi_t$ describes the flow of energy across part of the null cone; we will compute the integrand more explicitly in terms of the area form dK induced from an ambient *Euclidean* metric $dt^2 + \sum(dy^i)^2$, with the goal of showing that $\int_K \varphi_t \geq 0$. This area form is the contraction of the outward unit normal $\frac{1}{\sqrt{2}}(\frac{\partial}{\partial t} + \frac{y^i}{\|y\|}\frac{\partial}{\partial y^i})$ with the ambient Euclidean volume form $dt \wedge dy$, giving

$$dK = \tfrac{1}{\sqrt{2}}(dy - \tfrac{y^i}{\|y\|}dt \wedge dy_{(i)})|_K = \sqrt{2}\, dy|_K.$$

It is easy to calculate that the restriction to K of φ_t is

$$\varphi_t|_K = \tfrac{1}{\sqrt{2}}(e - \tfrac{y^i p_i}{\|y\|}p_0)dK.$$

Separating the radial and tangential space derivatives

$$p_r \stackrel{\text{def}}{=} \frac{y^i p_i}{\|y\|}, \quad p_\tau \stackrel{\text{def}}{=} \sqrt{\sum p_i^2 - p_r^2},$$

we can rewrite this as

$$\varphi_t|_K = \tfrac{1}{\sqrt{2}}(\tfrac{1}{2}(p_0^2 + p_r^2 + p_\tau^2) + F(z) - p_r p_0)dK \tag{3.72}$$

$$= \tfrac{1}{\sqrt{2}}(\tfrac{1}{2}(p_0 - p_r)^2 + \tfrac{1}{2}p_\tau^2) + F(z))dK. \tag{3.73}$$

In the region of $J^1(\mathbf{L}^{n+1}, \mathbf{R})$ where $F(z) \geq 0$ this integrand is positive, and from (3.71) we obtain the bound

$$\int_T e\,dy \leq \int_B e\,dy.$$

This says that "energy travels with at most unit speed"—no more energy can end up in T than was already present in B. If $F(z) \geq 0$ everywhere, then one can obtain another consequence of the expression (3.73) by writing

$$\int_K \varphi_t = \int_B \varphi_t - \int_T \varphi_t \leq \int_B \varphi_t \leq E(t_0) = E.$$

This gives an upper bound for

$$\|(dz|_K)\|_{L^2}^2 = \int_K ((p_0 - p_r)^2 + p_\tau^2)dK \leq 2\sqrt{2}E$$

which holds for the entire backward null cone; that is, our bound is independent of t_0. Here the L^2-norm is with respect to Euclidean measure.

We should also mention that the spatial-translation and Lorentz rotations give rise to conserved quantities that may be thought of as linear and angular momenta, respectively. The uses of these are similar to, though not as extensive as, the uses of the conserved energy.

3.5.2 The Conformally Invariant Wave Equation

We now determine some additional conservation laws for the conformally invariant wave equation (3.68). Again, we could duplicate the process used for Poisson equations by calculating restrictions of the right-invariant vector fields of $SO^o(n+1, 2)$ to the image of an embedding $J^1(\mathbf{L}^{n+1}, \mathbf{R}^+) \hookrightarrow SO^o(n+1, 2)$, and contracting with the left-invariant Lagrangian. Instead, we will illustrate the more concrete, coordinate-based approach, though we will still make some use of the geometry.

The Dilation Conservation Law

To find the conservation law corresponding to dilation symmetry of (3.68), we have to first determine a formula for this symmetry on $J^1(\mathbf{L}^{n+1}, \mathbf{R}^+)$, and then apply the Noether prescription. For this, we will first determine the vector field's action on $J^0(\mathbf{L}^{n+1}, \mathbf{R}^+)$; the lift of this action to $J^1(\mathbf{L}^{n+1}, \mathbf{R}^+)$ is determined by the requirement that it preserve the contact line bundle.

By analogy with (3.30), we have an embedding of $J^0(\mathbf{L}^{n+1}, \mathbf{R}^+)$ into the null-cone of $\mathbf{L}^{n+1,2}$ given by

$$(t, y^i, z) \mapsto z^{\frac{2}{n-1}} \begin{pmatrix} 1 \\ t \\ y^i \\ \frac{1}{2}(-t^2 + |y|^2) \end{pmatrix},$$

and the dilation matrix (in blocks of size $1, 1, n, 1$) acts projectively on this slice of the null-cone by

$$\left[z^{\frac{-2}{n-1}} \begin{pmatrix} 1 \\ t_r \\ y_r^i \\ \frac{1}{2}\|(t_r, y_r)\|^2) \end{pmatrix} \right] \overset{def}{=} \begin{pmatrix} r^{-1} & 0 & 0 & 0 \\ 0 & 1 & 0 & 0 \\ 0 & 0 & I_n & 0 \\ 0 & 0 & 0 & r \end{pmatrix} \cdot \left[z^{\frac{-2}{n-1}} \begin{pmatrix} 1 \\ t \\ y^i \\ \frac{1}{2}\|(t, y)\|^2 \end{pmatrix} \right].$$

Taking the derivative with respect to r and setting $r = 1$ gives the vector field

$$\bar{V}_{dil} = -\tfrac{n-1}{2} z \tfrac{\partial}{\partial z} + t \tfrac{\partial}{\partial t} + y^i \tfrac{\partial}{\partial y^i}.$$

The scaling in the z-coordinate reflects an interpretation of the unknown $z(t, y)$ as a section of a certain density line bundle. We then find the lift from $\bar{V}_{dil} \in \mathcal{V}(J^0(\mathbf{L}^{n+1}, \mathbf{R}^+))$ to $V_{dil} \in \mathcal{V}(J^1(\mathbf{L}^{n+1}, \mathbf{R}^+))$ by the requirement that the contact form $\theta = dz - p_0 dt - p_i dy^i$ be preserved up to scaling; that is,

$$V_{dil} = \bar{V}_{dil} + v^0 p_0 + v^i p_i$$

must satisfy

$$\mathcal{L}_{V_{dil}} \theta \equiv 0 \pmod{\{\theta\}},$$

where v^0 and v^i are the unknown coefficients of the lift. This simple calculation yields

$$V_{dil} = -\tfrac{n-1}{2} z \tfrac{\partial}{\partial z} + t \tfrac{\partial}{\partial t} + y^i \tfrac{\partial}{\partial y^i} - \tfrac{n+1}{2} p_a \tfrac{\partial}{\partial p_a}.$$

Then one can compute even for the general wave equation (3.67) that

$$\mathcal{L}_{V_{dil}} (L(t, y, z, p) dt \wedge dy) = ((n+1)F(z) - \tfrac{n-1}{2} z f(z)) dt \wedge dy.$$

Tentatively following the Noether prescription for the general wave equation, we set

$$\varphi_{dil} = V_{dil} \lrcorner \Lambda,$$

where Λ is given by (3.70), and because $\Lambda \equiv L\, dt \wedge dy$ modulo $\{I\}$, we can calculate

$$\begin{aligned} d\varphi_{dil} &= \mathcal{L}_{V_{dil}} \Lambda - V_{dil} \lrcorner \Pi \\ &\equiv \mathcal{L}_{V_{dil}} (L\, dt \wedge dy) - 0 \pmod{\mathcal{E}_\Lambda} \\ &= ((n+1)F(z) - \tfrac{n-1}{2} z f(z)) dt \wedge dy \pmod{\mathcal{E}_\Lambda}. \end{aligned}$$

The condition on the equation (3.67) that φ_{dil} be a conservation law is therefore

$$F(z) = C z^{\frac{2(n+1)}{n-1}},$$

so the PDE is

$$\Box z = C' z^{\frac{n+3}{n-1}},$$

as expected (cf. (3.68)); we work with $C' = 1$, $C = \frac{n-1}{2(n+1)}$. Now, one can calculate that restricted to any Legendre submanifold the conserved density is

$$\varphi_{dil} = L(t, y, z, p)(t\, dy - y^i dt \wedge dy_{(i)}) + (\tfrac{n-1}{2} z + t p_0 + y^i p_i)(p_0 dy + p_j dt \wedge dy_{(j)}).$$

Typically, one considers the restriction of this form to the constant-time hyperplanes $\mathbf{R}^n_t = \{t\} \times \mathbf{R}^n$, which is

$$\varphi_{dil} \equiv (te + r p_0 p_r + \tfrac{n-1}{2} z p_0)dy \quad (\text{mod } \{dt\}).$$

For example, we find that for solutions to (3.68) with compact support in y^i,

$$\frac{d}{dt} \int_{\mathbf{R}^n_t} (te + r p_0 p_r + \tfrac{n-1}{2} z p_0)dy = 0. \tag{3.74}$$

For more general wave equations (3.67), an identity like (3.74) holds, but with a non-zero right-hand side; our conservation law is a special case of this. The general dilation identity is of considerable use in the analysis of non-linear wave equations. It is analogous to the "almost-conservation law" derived from scaling symmetry used to obtain lower bounds on the area growth of minimal surfaces, as discussed in §1.4.3.

An Inversion Conservation Law

We now consider the inversion symmetry corresponding to the conjugate of time-translation by inversion in a unit (Minkowski) sphere. We will follow the same procedure as for dilation symmetry, first determining a vector field on $J^0(\mathbf{L}^{n+1}, \mathbf{R}^+)$ generating this inversion symmetry, then lifting it to a contact-preserving vector field on $J^1(\mathbf{L}^{n+1}, \mathbf{R}^+)$, and then applying the Noether prescription to obtain the conserved density.

The conjugate by sphere-inversion of a time-translation in $SO^o(n+1, 2)$ is the matrix

$$\begin{pmatrix} 1 & b & 0 & \tfrac{1}{2}b^2 \\ 0 & 1 & 0 & -b \\ 0 & 0 & I_n & 0 \\ 0 & 0 & 0 & 1 \end{pmatrix},$$

and differentiating its projective linear action on

$$\left[z^{-\frac{2}{n-1}} \begin{pmatrix} 1 \\ t \\ y^i \\ \tfrac{1}{2}\|(t,y)\|^2 \end{pmatrix} \right]$$

yields the vector field

$$\bar{V}_{inv} = \tfrac{n-1}{2}tz\tfrac{\partial}{\partial z} - \tfrac{1}{2}(t^2 + |y|^2)\tfrac{\partial}{\partial t} - ty^i\tfrac{\partial}{\partial y^i}.$$

Again, the coefficients of $\tfrac{\partial}{\partial t}$ and $\tfrac{\partial}{\partial y^i}$ describe an infinitesimal conformal motion of Minkowski space—representing an element of the Lie algebra of the Lorentz conformal group—and the coefficient of $\tfrac{\partial}{\partial z}$ gives the induced action on a density line bundle. We now look for coefficients $v_0\tfrac{\partial}{\partial p_0} + v_i\tfrac{\partial}{\partial p_i}$ to add to $\bar{V}_{inv}$ so that the new vector field will preserve θ up to scaling, and the unique solution is

$$V_{inv} = \bar{V}_{inv} + \left(\tfrac{n-1}{2}z + p_iy^i + \tfrac{n+1}{2}tp_0\right)\tfrac{\partial}{\partial p_0} + \left(p_0y^i + \tfrac{n+1}{2}tp_i\right)\tfrac{\partial}{\partial p_i}.$$

In applying the Noether prescription to Λ and V_{inv}, it will turn out that we need a compensating term, because $\mathcal{L}_{V_{inv}}\Lambda \not\equiv 0$ modulo $\mathcal{E}_\Lambda$. However, instead of performing this tedious calculation, we can simply test

$$
\begin{aligned}
\tilde{\varphi}_{inv} \ \overset{def}{=} \ & -V_{inv} \,\lrcorner\, \Lambda \\
\equiv \ & L(t,y,z,p)(\tfrac{1}{2}(t^2 + |y|^2)dy - ty^idt \wedge dy_{(i)}) \\
& + (\tfrac{n-1}{2}tz + \tfrac{1}{2}p_0(t^2 + |y|^2) + tp_iy^i) \wedge (p_0dy + p_jdt \wedge dy_{(j)}).
\end{aligned}
$$

We find that on solutions to (3.68),

$$d\tilde{\varphi}_{inv} = \tfrac{n-1}{2}p_0z\, dt \wedge dy = d(\tfrac{n-1}{4}z^2dy),$$

and we therefore set

$$\varphi_{inv} = \tilde{\varphi}_{inv} - \tfrac{n-1}{4}z^2dy.$$

As usual, we consider the restriction of this form to a hyperplane $\mathbf{R}^n_t = \{t\}\times\mathbf{R}^n$, which gives

$$\varphi_{inv} \equiv \left(\tfrac{1}{2}(L + p_0^2)(t^2 + |y|^2) + \tfrac{n-1}{2}p_0tz + tp_0p_iy^i - \tfrac{n-1}{4}z^2\right)dy,$$

modulo $\{dt, \theta\}$.

Again, for more general wave equations (3.67), this quantity gives not a conservation law, but a useful integral identity. The usefulness of the integrand follows largely from the fact that after adding an exact n-form, the coefficient of dy is positive. One notices this by expanding in terms of radial and tangential derivatives

$$
\begin{aligned}
\varphi_{inv} \ = \ & \Big\{\tfrac{1}{4}\left((p_0^2 + p_\tau^2 + p_r^2)(t^2 + r^2) + 2(n-1)p_0tz + 4trp_0p_r - (n-1)z^2\right) \\
& + \tfrac{1}{2}(t^2 + r^2)F(z)\Big\}\,dy,
\end{aligned}
$$

which suggests completing squares:

$$
\begin{aligned}
\varphi_{inv} \ = \ & \Big\{\tfrac{1}{4}\left(|p_0y + tp|^2 + (tp_0 + rp_r + (n-1)z)^2 + (rp_\tau)^2\right) \\
& + \tfrac{1}{2}(t^2 + r^2)F(z) - \tfrac{n-1}{4}(nz^2 + 2rzp_r)\Big\}\,dy.
\end{aligned}
$$

The last term is the divergence $d(\frac{n-1}{4}y^i z^2 dy_{(i)})$ modulo $\{dt, \theta\}$, and the remaining terms are positive. The positive expression

$$e_c \overset{def}{=} \tfrac{1}{2}\left(|p_0 y + t p|^2 + (t p_0 + r p_r + (n-1)z)^2 + (r p_\tau)^2\right) + (t^2 + r^2)F(z)$$

is sometimes called the *conformal energy* of the solution $z(t, y)$.

It is also sometimes convenient to set

$$e_d = r p_0 p_r + \tfrac{n-1}{2} z p_0$$

so that

$$\varphi_{dil} \equiv (e_d + te)dy \quad (\mathrm{mod}\ \{dt\}),$$

and also

$$\varphi_{inv} \equiv (t e_d + \tfrac{1}{2}(t^2 + |y|^2)e - \tfrac{n-1}{4}z^2)dy \quad (\mathrm{mod}\ \{dt\}).$$

The fact that the integrals of these quantities are constant in t yields results about growth of solutions.

In fact, in the analysis of wave equations that are perturbations of the conformally invariant equation (3.68), the most effective estimates pertain to the quantities appearing in these conservation laws. One can think of the conservation laws as holding for our "flat" non-linear wave equation (3.68), and then the estimates are their analogs in the "curved" setting.

3.5.3   Energy in Three Space Dimensions

We conclude this section by discussing a few more properties that involve the energy in three space dimensions.

The fact that the energy $E(t)$ is constant implies in particular that $\int_{\mathbf{R}^n_t} z_t dy$ is bounded with respect to t. This allows us to consider the evolution of the spatial L^2-norm of a solution to $\Box z = z^3$ as follows:

$$
\begin{aligned}
\frac{d^2}{dt^2}\left(\int_{\mathbf{R}^3_t} z^2 dy\right) &= 2\int_{\mathbf{R}^3_t}(z\, z_{tt} dy + z_t^2)dy \\
&= 2\int_{\mathbf{R}^3_t}(z\Delta z - z^4 + z_t^2)dy \\
&= -2\int_{\mathbf{R}^3_t}(|\nabla_y z|^2 + z^4)dy + 2\int_{\mathbf{R}^n_t} z_t^2 dy \\
&\leq 4E,
\end{aligned}
$$

where the second equality follows from the differential equation and the third from Green's theorem (integration by parts). The conclusion is that $\|z\|^2_{L^2(\mathbf{R}^n_t)}$ grows at most quadratically,

$$\int_{\mathbf{R}^3_t} z^2 dy \leq 2Et^2 + C_1 t + C_2, \tag{3.75}$$

and in particular $\int z^2 dy = O(t^2)$.

The energy plays another interesting role in the equation

$$\Box z = -z^3 \tag{3.76}$$

Here, it is possible for the energy to be negative:

$$E = \int_{\mathbf{R}_t^3} \left(\tfrac{1}{2}(z_t^2 + ||\nabla z||^2) - \tfrac{z^4}{4} \right) dy. \tag{3.77}$$

We will prove that a solution to (3.76), with compactly supported initial data $z(0, y)$, $z_t(0, y)$ satisfying $E < 0$, must blow up in finite time ([Lev74]). Notice that any non-trivial compactly supported initial data may be scaled up to achieve $E < 0$, and may be scaled down to achieve $E > 0$.

The idea is to show that the quantity

$$I(t) \stackrel{def}{=} \int_{\mathbf{R}_t^3} \tfrac{1}{2} z^2 dy$$

becomes unbounded as $t \nearrow T$ for some finite time $T > 0$. We start by computing its derivatives

$$
\begin{aligned}
I'(t) &= \int z z_t dy, \\
I''(t) &= \int z_t^2 dy + \int z z_{tt} dy \\
&= \int z_t^2 dy - \int |\nabla z|^2 dy + \int z^4 dy.
\end{aligned}
$$

The last step uses Green's theorem, requiring the solution to have compact y-support for each $t \geq 0$. To dispose of the $\int z^4$ term, we add $4E$ to each side using (3.77):

$$I''(t) + 4E = 3 \int z_t^2 dy + \int |\nabla z|^2 dy.$$

We can discard from the right-hand side the positive gradient term, and from the left-hand side the negative energy term, to obtain

$$I''(t) > 3 \int z_t^2 dy.$$

To obtain a second-order differential inequality for I, we multiply the last inequality by $I(t)$ to obtain

$$
\begin{aligned}
I(t)I''(t) &> \frac{3}{2} \left(\int z^2 dy \right) \left(\int z_t^2 dy \right) \\
&\geq \tfrac{3}{2} I'(t)^2.
\end{aligned}
$$

The last step follows from the Cauchy-Schwarz inequality, and says that $I(t)^{-1/2}$ has negative second derivative. We would like to use this to conclude that $I^{-1/2}$ vanishes for some $T > 0$ (which would imply that I blows up), but for this we would need to know that $(I^{-1/2})'(0) < 0$, or equivalently $I'(0) > 0$, which may not hold.

To rectify this, we shift I to

$$J(t) = I(t) - \tfrac{1}{2}E(t+\tau)^2,$$

with $\tau > 0$ chosen so that $J'(0) > 0$. We now mimic the previous reasoning to show that $(J^{-1/2})''(t) < 0$. We have

$$J'(t) = \int zz_t dy - E(t+\tau),$$

$$J''(t) = \int z_t^2 dy + \int zz_{tt} dy - E$$

$$= 3\int z_t^2 dy + \int \|\nabla z\|^2 dy - 5E$$

$$> 3\left(\int z_t^2 dy - E\right).$$

From this we obtain

$$J(t)J''(t) - \tfrac{3}{2}J'(t)^2 > \tfrac{3}{2}\left[\left(\int z^2 - E(t+\tau)^2\right)\left(\int z_t^2 - E\right) - \left(\int zz_t - E(t+\tau)\right)^2\right],$$

which is positive, again by the Cauchy-Schwarz inequality. This means that $(J^{-1/2})''(t) < 0$. Along with $J^{-1/2}(0) > 0$ and $(J^{-1/2})'(0) < 0$, this implies that for some $T > 0$, $J^{-1/2}(T) = 0$, so $J(t)$ blows up.

We conclude by noting that the qualitative behavior of solutions of $\Box z = f(z)$ depends quite sensitively on the choice of non-linear term $f(z)$. In contrast to the results for $\Box z = \pm z^3$ described above, we have for the equation

$$\Box z = -z^2 \qquad (n = 3),$$

that every solution must blow up in finite time ([Joh79]). We will outline the proof in case the initial data are compactly supported and satisfy

$$\int u(0,t)dy > 0, \qquad \int u_t(0,t)dy > 0.$$

Note that replacing z by $-z$ gives the equation $\Box z = z^2$, which therefore has the same behavior.

This proof is fairly similar to the previous one; we will derive differential inequalities for

$$J(t) \overset{\text{def}}{=} \int_{\mathbf{R}_t^n} z\, dy$$

which imply that this quantity blows up. We use integration by parts to obtain

$$J''(t) = \int z_{tt} dy = \int (\Delta z + z^2) dy = \int z^2,$$

and using Hölder's inequality on Supp $z \subset \{|y| \leq R_0 + t\}$ in the form $||z||_{L^1} \leq ||z||_{L^2} ||1||_{L^2}$, this gives

$$J''(t) \geq C \left(\int z\, dy \right)^2 (R_0 + t)^{-3} \geq C(1+t)^{-3} J(t)^2. \qquad (3.78)$$

This is the first ingredient.

Next, we use the fact that if $z_0(y, t)$ is the *free solution* to the homogeneous wave equation $\Box z_0 = 0$, with the same initial data as our z, then

$$z(y, t) \geq z_0(y, t)$$

for $t \geq 0$; this follows from a certain explicit integral expression for the solution. Note that if we set $J_0(t) = \int_{\mathbf{R}^n} z_0$, then it follows from the equation alone that $J_0''(t) = 0$, and by the hypotheses on our initial data we have $J_0(t) = C_0 + C_1 t$ with $C_0, C_1 > 0$. Another property of the free solution is that its support at time t lies in the annulus $A_t = \{t - R_0 \leq |y| \leq t + R_0\}$. Now

$$
\begin{aligned}
C_0 + C_1 t &\leq \int_{A_t} z\, dy \\
&\leq ||z||_{L^1(A)} \\
&\leq ||z||_{L^2(A)} ||1||_{L^2(A)} \\
&\leq C(1+t) \left(\int z^2 dy \right)^2.
\end{aligned}
$$

This gives

$$J''(t) = \int z^2 dy \geq \left(\frac{C_0 + C_1 t}{C(1+t)} \right)^{1/2},$$

and in particular, $J'' > 0$. With the assumptions on the initial data, this gives

$$
\begin{aligned}
J'(t) &> 0 \\
J(t) &\geq C(1+t)^2.
\end{aligned}
$$

We can use (3.78, 3.79, 3.79) to conclude that J must blow up at some finite time. This follows by writing

$$
\begin{aligned}
J''(t) &\geq C(1+t)^{-3} J^{3/2} J^{1/2} \\
&\geq C(1+t)^{-2} J^{3/2}.
\end{aligned}
$$

Multiply by J' and integrate to obtain

$$J'(t) \geq C(1+t)^{-1} J(t)^{5/4}.$$

Integrating again, we have

$$J(t) \geq \left(J(0)^{-1/4} - \tfrac{1}{4}C\ln(1+t)\right)^{-4},$$

and because $J(0) > 0$ and $C > 0$, $J(t)$ must blow up in finite time.

Chapter 4

Additional Topics

4.1 The Second Variation

In this section, we will discuss the second variation of the Lagrangian functionals considered in the preceding chapters. We begin by giving an invariant, coordinate-free calculation of the formula (4.8) for the second derivative of a functional under fixed-boundary variations. This formula has an interpretation in terms of conformal structures induced on integral manifolds of the Euler-Lagrange system, which we will describe. The role played by conformal geometry here is not to be confused with that in the previous chapter, although both situations seem to reflect the increasing importance of variational equations in conformal geometry.

The usual integration by parts that one uses to establish local minimality of a solution to the Euler-Lagrange equations cannot generally be done in an invariant manner, and we discuss a condition under which this difficulty can be overcome. We considered in §2.5 the example of prescribed mean curvature hypersurfaces in Euclidean space; we will give an invariant calculation of the second variation formula and the integration by parts for this example. We conclude by discussing various classical conditions under which an integral manifold of an Euler-Lagrange system is locally minimizing, using the Poincaré-Cartan form to express and prove some of these results in a coordinate-free manner.

4.1.1 A Formula for the Second Variation

We start by reconsidering the situation of §1.2, in which we calculated the first variation of a Lagrangian Λ on a contact manifold. This amounted to taking the first derivative, at some fixed time, of the values of the functional $\mathcal{F}_\Lambda$ on a 1-parameter family of Legendre submanifolds. Our goal is to extend the calculation to give the second variation of Λ, or equivalently, the second derivative of $\mathcal{F}_\Lambda$ on a 1-parameter family at a Legendre submanifold for which the first variation vanishes. The result appears in formula (4.5) below, and in a more geometric form in (4.8). This process is formally analogous to computing the Hessian matrix of a smooth function $f : \mathbf{R}^n \to \mathbf{R}$ at a critical point, which is typically done with the goal of identifying local extrema.

Let (M, I) be a contact manifold, with contact form $\theta \in \Gamma(I)$, and Lagrangian $\Lambda \in \Omega^n(M)$ normalized so that the Poincaré-Cartan form is given by $\Pi = d\Lambda = \theta \wedge \Psi$. Fix a compact manifold N^n with boundary ∂N, and a smooth map

$$F : N \times [0, 1] \to M$$

which is a Legendre submanifold F_t for each fixed $t \in [0, 1]$ and is independent of t on $\partial N \times [0, 1]$. Two observations will be important:

- $F^*\theta = G\, dt$ for some function G on $N \times [0, 1]$, depending on the choice of generator $\theta \in \Gamma(I)$. This holds because each F_t is a Legendre submanifold, meaning that $F_t^*\theta = 0$.

- For every form $\alpha \in \Omega^*(M)$, and every boundary point $p \in \partial N$, we have

$$\left(\tfrac{\partial}{\partial t} \lrcorner F^*\alpha \right)(p, t) = 0.$$

This is equivalent to the fixed-boundary condition; at each $p \in \partial N$, we have $F_*(\tfrac{\partial}{\partial t}) = 0$.

We previously calculated the first variation as (see §1.2.2)

$$\frac{d}{dt} \left(\int_{N_t} F_t^*\Lambda \right) = \int_{N_t} G \cdot F_t^*\Psi,$$

where $\Pi = d\Lambda = \theta \wedge \Psi$ is the Poincaré-Cartan form for Λ. This holds for each $t \in [0, 1]$.

We now assume that F_0 is stationary for Λ; that is, F is a Legendre variation of an integral manifold $F_0 : N \hookrightarrow M$ of the Euler-Lagrange system $\mathcal{E}_\Lambda = \{\theta, d\theta, \Psi\}$. This is the situation in which we want to calculate the second derivative:

$$
\begin{aligned}
\delta^2(\mathcal{F}_\Lambda)_{N_0}(g) &= \frac{d^2}{dt^2}\bigg|_{t=0} \left(\int_{N_t} F_t^*\Lambda \right) \\
&= \frac{d}{dt}\bigg|_{t=0} \int_{N_t} G\, F_t^*\Psi \\
&= \int_{N_0} \mathcal{L}_{\frac{\partial}{\partial t}} (G\, F^*\Psi) \\
&= \int_{N_0} g\, \mathcal{L}_{\frac{\partial}{\partial t}} (F^*\Psi),
\end{aligned}
$$

where $g = G|_{t=0}$, and the last step uses the fact that $F_0^*\Psi = 0$.

To better understand the Lie derivative $\mathcal{L}_{\frac{\partial}{\partial t}}(F^*\Psi)$, we use the results obtained via the equivalence method in §2.4. This means that we are restricting our attention to the case $n \geq 3$, with a Poincaré-Cartan form that is neo-classical and definite. In this situation, we have a G-structure $B \to M$, where $G \subset GL(2n+1, \mathbf{R})$ has Lie algebra

$$\mathfrak{g} = \left\{ \begin{pmatrix} (n-2)r & 0 & 0 \\ 0 & -2r\delta^i_j + a^i_j & 0 \\ d_i & s_{ij} & nr\delta^j_i - a^j_i \end{pmatrix} : a^i_j + a^j_i = s_{ij} - s_{ji} = s_{ii} = 0 \right\}. \tag{4.1}$$

The sections of $B \to M$ are local coframings $(\theta, \omega^i, \pi_i)$ of M for which:

- θ generates the contact line bundle I,

- the Poincaré-Cartan form is $\Pi = -\theta \wedge \pi_i \wedge \omega_{(i)}$,

- there exists a $\mathfrak{g}$-valued 1-form

$$\varphi = \begin{pmatrix} (n-2)\rho & 0 & 0 \\ 0 & -2\rho\delta^i_j + \alpha^i_j & 0 \\ \delta_i & \sigma_{ij} & n\rho\delta^j_i - \alpha^j_i \end{pmatrix}, \tag{4.2}$$

satisfying a structure equation

$$d \begin{pmatrix} \theta \\ \omega^i \\ \pi_i \end{pmatrix} = -\varphi \wedge \begin{pmatrix} \theta \\ \omega^j \\ \pi_j \end{pmatrix} + \begin{pmatrix} -\pi_i \wedge \omega^i \\ \Omega^i \\ 0 \end{pmatrix} \tag{4.3}$$

where

$$\Omega^i = T^{ijk}\pi_j \wedge \omega^k - (S^i_j\omega^j + U^{ij}\pi_j) \wedge \theta, \tag{4.4}$$

with $T^{ijk} = T^{jik} = T^{kji}$, $T^{iik} = 0$; $U^{ij} = U^{ji}$; $S^i_j = S^j_i$, $S^i_i = 0$. The pseudo-connection form φ may be chosen so that also (cf. (2.48))

$$(n-2)d\rho = -\delta_i \wedge \omega^i - S^i_j\pi_i \wedge \omega^j + \left(\tfrac{n-2}{2n}U^{ij}\sigma_{ij} - t^i\pi_i\right) \wedge \theta$$

for some functions t^i.

For any point $p \in N$, we consider a neighborhood $U \subset M$ of $F_0(p)$ on which we can fix one such coframing $(\theta, \omega^i, \pi_i)$ with pseudo-connection φ. All of the forms and functions may be pulled back to $W = F^{-1}(U) \subset N \times [0, 1]$, which is the setting for the calculation of $\mathcal{L}_{\frac{\partial}{\partial t}}(F^*\Psi)$. From now on, we drop all F^*s.

We have $\Psi = -\pi_i \wedge \omega_{(i)}$, and we now have the structure equations needed to differentiate Ψ, but it will simplify matters if we further adapt the forms $(\theta, \omega^i, \pi_i)$ on W in a way that does not alter the structure equations. Note first that restricted to each $W_t = W \cap N_t$, we have $\bigwedge \omega^i \neq 0$, so $(\omega^1, \dots, \omega^n, dt)$ forms a coframing on W. We can therefore write $\pi_i = s_{ij}\omega^j + g_i dt$ for some s_{ij}, g_i, and because each W_t is Legendre, we must have $s_{ij} = s_{ji}$. The structure group of $B \to M$ admits addition of a traceless, symmetric combination of the ω^j to the π_i, so we replace

$$\pi_i \rightsquigarrow \pi_i - s^o_{ij}\omega^j,$$

where $s_{ij}^0 = s_{ij} - \frac{1}{n}\delta_j^i s_{kk}$ is the traceless part. Now we have

$$\pi_i = s\omega^i + q_i dt$$

for some functions s, q_i on W, so that

$$\Psi = -\pi_i \wedge \omega_{(i)} = -ns\,\omega - q_i dt \wedge \omega_{(i)}.$$

Note that because F_0 is assumed integral for the Euler-Lagrange system, we have $s = 0$ everywhere on $W_0 \subset W$; in particular, $\pi_i|_{W_0} = 0$.

With our choice of π_i, recalling that along W_0, $\theta = g\,dt$ for some function g on W_0, and keeping in mind that restricted to W_0 we have $\pi_i = 0$ and $s = 0$, we can calculate

$$
\begin{aligned}
\mathcal{L}_{\frac{\partial}{\partial t}}\Psi\Big|_{t=0} &= -\tfrac{\partial}{\partial t} \lrcorner\, d(\pi_i \wedge \omega_{(i)}) - d(\tfrac{\partial}{\partial t} \lrcorner\,(\pi_i \wedge \omega_{(i)})) \\
&= \tfrac{\partial}{\partial t} \lrcorner\, \Big((\delta_i \wedge \theta + (n\rho\delta_i^j - \alpha_i^j) \wedge \pi_j) \wedge \omega_{(i)} \\
&\qquad + (s\omega^i + q_i dt) \wedge d\omega_{(i)} \Big) \\
&\qquad - d\big((q_i + s\tfrac{\partial}{\partial t} \lrcorner\, \omega^i)\omega_{(i)} - \pi_i \wedge (\tfrac{\partial}{\partial t} \lrcorner\, \omega_{(i)}) \big) \\
&= -(g\delta_i + dq_i + n\rho q_i - q_j \alpha_i^j) \wedge \omega_{(i)}.
\end{aligned}
$$

This gives our desired formula:

$$\frac{d^2}{dt^2}\Big|_{t=0} \left(\int_{N_t} \Lambda \right) = -\int_{N_0} g(dq_i + n\rho q_i - q_j\alpha_i^j + g\delta_i) \wedge \omega_{(i)}. \tag{4.5}$$

Unfortunately, in its present form this is not very enlightening, and our next task is to give a geometric interpretation of the formula.

4.1.2   Relative Conformal Geometry

It is natural to ask what kind of geometric structure is induced on an integral manifold $f : N \hookrightarrow M$ of an Euler-Lagrange system $\mathcal{E}_\Lambda$. What we find is:

> If $\Pi = d\Lambda$ is a definite, neo-classical Poincaré-Cartan form, then an integral manifold N of its Euler-Lagrange system $\mathcal{E}_\Lambda$ has a natural conformal structure, invariant under symmetries of (M, Π, N), even though there may be no invariant conformal structure on the ambient M.

This is a simple pointwise phenomenon, in the sense that any n-plane $V^n \subset T_p M^{2n+1}$ on which $\bigwedge \omega^i \neq 0$ has a canonical conformal inner-product defined as follows. Taking any section $(\theta, \omega^i, \pi_i)$ of $B \to M$, we can restrict the induced quadratic form $\sum(\omega^i)^2$ on TM to $V \subset T_p M$, where it is positive definite, and then the action of the structure group (4.1) on (ω^i) shows that up to scaling, this quadratic form is independent of our choice of section. Alternatively, one can show this infinitesimally by using the structure equations to compute on B

the Lie derivative of $\sum(\omega^i)^2$ along a vector field that is vertical for $B \to M$; this Lie derivative is itself multiple of $\sum(\omega^i)^2$. Note that we have *not* restricted to the conformal branch of the equivalence problem, characterized by $T^{ijk} = U^{ij} = S^i_j = 0$ and discussed in §3.3.

In particular, any integral manifold $f : N \hookrightarrow M$ for the Euler-Lagrange system $\mathcal{E}_\Lambda$ inherits a canonical conformal structure $[ds^2]_f$. We now want to develop the conformal structure equations for $(N, [ds^2]_f)$, in terms of the structure equations on $B \to M$, and our procedure will work only for integral manifolds of $\mathcal{E}_\Lambda$. We first note that along our integral manifold N we can choose local sections $(\theta, \omega^i, \pi_i)$ of B_N which are adapted to N in the sense that

$$T_p N = \{\theta, \pi_1, \cdots, \pi_n\}^\perp \subset T_p M,$$

for each $p \in N$. In fact, such sections define a reduction $B_f \to N$ of the principal bundle $B_N \to N$, having Lie algebra defined as in (4.1) by $s_{ij} = 0$.

Restricted to B_f, we have the same structure equations as on B, but with $\theta = \pi_i = 0$, and $d\theta = d\pi_i = 0$. Now observe that two of our structure equations restrict to give

$$d\omega^i = (2\rho\delta^i_j - \alpha^i_j) \wedge \omega^j,$$
$$d\rho = -\tfrac{1}{n-2}\delta_i \wedge \omega^i.$$

We therefore have a situation similar to that in §3.3 (cf. (3.40, 3.41)), with equations formally like those in the conformal geometry equivalence problem of §3.1.2. We can mimic the derivation of conformal structure equations in the present case by first setting $\beta_i = \frac{2}{n-2}\delta_i$, and then we know that this results in an equation

$$d\alpha^i_j + \alpha^i_k \wedge \alpha^k_j + \beta_i \wedge \omega^j - \beta_j \wedge \omega^i = \tfrac{1}{2}A^i_{jkl}\omega^k \wedge \omega^l, \tag{4.6}$$

where the quantity (A^i_{jkl}) has the symmetries of the Riemann curvature tensor. Furthermore, we know that there are unique functions $t_{ij} = t_{ji}$ such that replacing $\beta_i \rightsquigarrow \beta_i + t_{ij}\omega^j$ will yield the preceding equation with $A^i_{jkl} = 0$. However, it will simplify matters later if we go back and replace instead

$$\delta_i \rightsquigarrow \delta_i + \tfrac{n-2}{2}t^o_{ij}\omega^j,$$

where $t^o_{ij} = t_{ij} - \tfrac{1}{n}\delta_{ij}t_{kk}$ is the traceless part (note that only a *traceless* addition to δ_i will preserve the structure equations on $B \to M$). In terms of the new δ_i, we define

$$\beta_i \stackrel{def}{=} \tfrac{2}{n-2}(\delta_i - \tfrac{1}{n}K\omega^i), \tag{4.7}$$

where

$$K \stackrel{def}{=} -\tfrac{n-2}{2}t_{jj}.$$

This K was chosen so that defining β_i by (4.7) gives the correct conformal structure equation (4.6) with $A^l_{jkl} = 0$; it reflects the difference between the pseudo-connection forms for the Poincaré-Cartan form and the Cartan connection forms for the induced conformal geometry on the submanifold. One interpretation is the following.

> *The function K on B_f is a fundamental invariant of a stationary submanifold $(N, [ds^2]_f) \hookrightarrow (M, \mathrm{II})$ of an Euler-Lagrange system, and may be thought of as an extrinsic curvature depending on up to third derivatives of the immersion f.*

In the classical setting, N is already the 1-jet graph of a solution $z(x)$ of an Euler-Lagrange equation, so an expression for $K(x)$ depends on fourth derivatives of $z(x)$.

Now suppose that our integral manifold $f : N \hookrightarrow M$, with δ_i, β_i, and K as above, is the initial manifold $F_0 = f$ in a Legendre variation $F : N \times [0, 1] \to M$. Then we can rewrite our formula (4.5) for the second variation as

$$\delta^2(\mathcal{F}_\Lambda)_{N_0}(g) = - \int_{N_0} \left(g(dg_i + n\rho g_i - g_j \alpha_i^j + \tfrac{n-2}{2} g\beta_i) \wedge \omega_{(i)} + g^2 K\omega \right).$$

Part of this integrand closely resembles the expression (3.19) for the second covariant derivative of a section of a density line bundle, discussed in constructing the conformal Laplacian in §3.1.3. This suggests the following computations.

First, consider the structure equation $d\theta = -(n-2)\rho \wedge \theta - \pi_i \wedge \omega^i$. Along W_0 (but not yet restricted to W_0), where $\theta = g\,dt$ and $\pi_i = g_i dt$, this reads

$$dg \wedge dt = -(n-2)\rho \wedge g\,dt - g_i dt \wedge \omega^i,$$

so that restricted to W_0, we have

$$dg + (n-2)\rho g = g_i \omega^i.$$

This equation should be interpreted on B_f, which is identified with the principal bundle associated to the conformal structure $[ds^2]_f$. It says that g is a section of the density bundle $D^{\frac{n-2}{2n}}$, and that g_i are the components of its covariant derivative (see (3.18) ff.). We can now write

$$dg_i + n\rho g_i - g_j \alpha_i^j + \tfrac{n-2}{2} g\beta_i = g_{ij}\omega^j,$$

and by definition

$$\Delta_f g = g_{ii} \in \Gamma(D^{\frac{n+2}{2n}}),$$

where Δ_f is the conformal Laplacian on N induced by $f = F_0 : N \hookrightarrow M$. We now have a more promising version of the second variation formula:

$$\boxed{\delta^2(\mathcal{F}_\Lambda)_{N_0}(g) = - \int_{N_0} (g\Delta_f g + g^2 K)\omega.} \tag{4.8}$$

It is worth noting that the sign of this integrand does not depend on the sign of the variation's generating function g, and if we fix an orientation of N, then the integrand $K\omega$ on B_f has a well-defined sign at each point of N.

4.1.3 Intrinsic Integration by Parts

In order to detect local minima using the second variation formula (4.8), it is often helpful to convert an integral like $\int g\Delta g\,dx$ into one like $-\int \|\nabla g\|^2 dx$. In the Euclidean setting, with either compact supports or with boundary terms, this is done with integration by parts and is straightforward; the two integrands differ by the divergence of $g\nabla g$, whose integral depends only on boundary data.

We would like to perform a calculation like this on N, for an arbitrary Legendre variation g, using only intrinsic data. In other words, we would like to associate to any N and g some $\xi \in \Omega^{n-1}(N)$ such that $d\xi$ is the difference between $(g\Delta_f g)\omega$ and some quadratic expression $Q(\nabla g, \nabla g)\omega$, possibly with some additional zero-order terms. A natural expression to consider, motivated by the flat case, is

$$\xi = gg_i\omega_{(i)}. \tag{4.9}$$

Here, in order for g_i to make sense, we are assuming that we have a local coframing on M adapted to the integral manifold N, so that at points of N, $\theta = g\,dt$, $\pi_i = g_i dt$, and so that restricted to N, $\theta|_N = \pi_i|_N = 0$. We can then compute:

$$
\begin{aligned}
d\xi &= dg \wedge g_i\omega_{(i)} + g\,dg_i \wedge \omega_{(i)} + gg_i d\omega^j \wedge \omega_{(ij)} &\tag{4.10}\\
&= (dg + (n-2)\rho g) \wedge g_i\omega_{(i)} + g(dg_i + n\rho g_i - g_j\alpha_i^j) \wedge \omega_{(i)} &\tag{4.11}\\
&= \left(\sum(g_i)^2 + g\Delta_f g\right)\omega - \tfrac{n-2}{2}g^2\beta_i \wedge \omega_{(i)}. &\tag{4.12}
\end{aligned}
$$

Now, the first term is exactly what we are looking for, and second is fairly harmless because it is of order zero in the variation g, and in practice contributes only terms similar to $g^2 K\omega$.

The problem is that ξ (4.9) is defined on the total space $B_f \to N$, and although semibasic for this bundle, it is not basic; that is, there is no form on N that pulls back to B_f to equal ξ, even locally. The criterion for ξ to be basic is that $d\xi$ be semibasic, and this fails because of the appearance of β_i in (4.12).

But suppose that we can find some canonical reduction of the ambient $B \to M$ to a subbundle on which δ_i becomes semibasic over M; in terms of the Lie algebra (4.1), this means that we can reduce to the subgroup having Lie algebra given by $\{d_i = 0\}$. In this case, each $\beta_i = \frac{2}{n-2}(\delta_i - \frac{K}{n}\omega^i)$ is a linear combination of θ, ω^i, π_i, and is in particular semibasic over N. Consequently, ξ is basic over N, and we can perform the integration by parts in an invariant manner.

Unfortunately, there are cases in which no such canonical reduction of B is possible. An example is the homogeneous Laplace equation on $\mathbf{R}^n$,

$$\Delta z = 0,$$

which is preserved under an action of the conformal group $SO^\circ(n+1, 1)$. The associated conformal geometry on the trivial solution $z = 0$ is flat, and our second variation formula reads

$$\delta^2(\mathcal{F}_\Lambda)_0(g) = \int_\Omega g\Delta g\,dx$$

for a variation $g \in C_0^\infty(\Omega)$, $\Omega \subset \mathbf{R}^n$. It follows from our construction that this integrand is invariant under a suitable action of the conformal group. However, the tempting integration by parts

$$\int_\Omega g \Delta g \, dx = -\int_\Omega \|\nabla g\|^2 dx$$

leaves us with an integrand which is *not* conformally invariant. It is this phenomenon that we would like to avoid.

To get a sense of when one might be able to find a canonical subbundle of $B \to M$ on which the δ^i are semibasic, recall that

$$(n-2)d\rho \equiv -\delta_i \wedge \omega^i \quad (\text{mod } \{\theta, \pi_i\}).$$

Working modulo $\{\theta, \pi_i\}$ essentially amounts to restricting to integral manifolds of the Euler-Lagrange system. The preceding then says that a choice of subbundle $B' \subset B$ on which δ_i are semibasic gives a subbundle of each conformal bundle $B_f \to N$ on which $d\rho$ is semibasic over N. Now, typically the role of ρ in the Cartan connection for a conformal structure is as a pseudo-connection in the density line bundle D; a special reduction is required for ρ to be a genuine connection, and the latter requirement is equivalent to $d\rho$ being semibasic. In other words, being able to integrate by parts in an invariant manner as described above is equivalent to having a connection in D represented by the pseudo-connection ρ. One way to find a connection in D is to suppose that D has somehow been trivialized, and this is equivalent to choosing a Riemannian metric representing the conformal class. This suggests that Euler-Lagrange systems whose integral manifolds have canonical Riemannian metrics will have canonical reductions of this type.

In fact, we have seen an example of a Poincaré-Cartan form whose geometry $B \to M$ displays this behavior. This is the system for Riemannian hypersurfaces having prescribed mean curvature, characterized in §2.5 in terms of differential invariants of its neo-classical, definite Poincaré-Cartan form. To illustrate the preceding discussion, we calculate the second variation formula for this system. The reader should note especially how use of the geometry of the Poincaré-Cartan form gives a somewhat simpler derivation of the formula than one finds in standard sources.[1]

4.1.4   Prescribed Mean Curvature, Revisited

In §2.5, we considered a definite, neo-classical Poincaré-Cartan form (M, Π) whose associated geometry $(B \to M, \varphi)$ had invariants satisfying

$$T^{ijk} = 0, \quad U^{ij} = \lambda \delta^{ij}.$$

We further assumed the open condition

$$\lambda < 0,$$

---

[1] See for example pp. 513–539 of [Spi75], where the calculation is prefaced by a colorful warning about its difficulty.

and this led to a series of reductions of $B \to M$, resulting in a principal sub-bundle $B_3 \to M$, having structure group with Lie algebra

$$\mathfrak{g}_3 = \left\{ \begin{pmatrix} 0 & 0 & 0 \\ 0 & a^i_j & 0 \\ 0 & 0 & -a^j_i \end{pmatrix} : a^i_j + a^j_i = 0 \right\},$$

on which the original structure equations (4.2, 4.3, 4.4) hold, with $U^{ij} = -\delta^{ij}$, $S^i_j = T^{ijk} = 0$, $\rho = -\frac{H}{2n}\theta$ for a function H on B_3, and $d\rho = -\frac{1}{n-2}\delta_i \wedge \omega^j$, where $\delta_i \equiv 0 \pmod{\{\theta, \omega^j, \pi_j\}}$. In this case we computed that $dH \equiv 0 \pmod{\{\theta, \omega^i\}}$, so that H defines a function on the local quotient space Q^{n+1}, which also inherits a Riemannian metric $\sum(\omega^i)^2$. The contact manifold M can be locally identified with the bundle of tangent hyperplanes of Q, and the integral manifolds of the Euler-Lagrange differential system $\mathcal{E}_\Lambda$ are the tangent loci of hypersurfaces in Q whose mean curvature coincides with the background function H. In this case $B_3 \to M \to Q$ is locally identified with the orthonormal frame bundle of the Riemannian manifold Q.

An important point here is that the Riemannian geometry associated to Π appears only after reducing to $B_3 \to M$. However, if our goal is to see the formula for the second variation, then we face the following difficulty. That formula required the use of coframes of M adapted to a stationary submanifold $N \hookrightarrow M$ in a certain way, but while adapted coframes can always be found in $B \to M$, there is no guarantee that they can be found in the subbundle $B_3 \to M$, where the Riemannian geometry is visible.

We will overcome these difficulties and illustrate the invariant calculation of the second variation by starting only with the Riemannian geometry of (Q, ds^2). This is expressed in the Levi-Civita connection in the orthonormal frame bundle, where we can also give the Poincaré-Cartan form and Euler-Lagrange system for prescribed mean curvature. We then introduce higher-order data on a larger bundle, which allows us to study the second fundamental form. In fact, this larger bundle corresponds to the partial reduction $B_2 \to M$ on which ρ and δ_i are semibasic, but σ_{ij} is not. The end result of our calculation is formula (4.16). In the following discussion, index ranges are $0 \le a, b, c \le n$ and $1 \le i, j, k \le n$.

We begin with a generalization of the discussion in §1.4 of constant mean curvature hypersurfaces in Euclidean space. Let (Q, ds^2) be an oriented Riemannian manifold of dimension $n + 1$. A *frame* for Q is a pair $f = (q, e)$ consisting of a point $q \in Q$ and a positively oriented orthonormal basis $e = (e_0, \dots, e_n)$ for $T_q Q$. The set $\mathcal{F}$ of all such frames is a manifold, and the right $SO(n + 1, \mathbf{R})$-action

$$(q, (e_0, \dots, e_n)) \cdot (g^a_b) = (q, (\sum e_a g^a_0, \dots, \sum e_a g^a_n))$$

gives the basepoint map

$$q : \mathcal{F} \to Q$$

the structure of a principal bundle. The unit sphere bundle

$$M^{2n+1} = \{(q, e_0) : q \in Q, \ e_0 \in T_q Q, \ \|e_0\| = 1\}$$

is identified with the Grassmannian bundle of oriented tangent n-planes in TQ, and it has a contact structure generated by the 1-form

$$\theta_{(q,e_0)}(v) = ds^2(e_0, q_*(v)), \qquad v \in T_{(q,e_0)}M, \tag{4.13}$$

where $q : M \to Q$ is the projection. An immersed oriented hypersurface in Q has a unit normal vector field, which may be thought of as a 1-jet lift of the submanifold to M. The submanifold of M thus obtained is easily seen to be a Legendre submanifold for this contact structure, and the transverse Legendre submanifold is locally of this form.

To carry out calculations on M, and even to verify the non-degeneracy of θ, we will use the projection $\mathcal{F} \to M$ defined by $(q, (e_0, \ldots, e_n)) \mapsto (q, e_0)$. Calculations can then be carried out using structure equations for the canonical parallelization of $\mathcal{F}$, which we now introduce. First, there are the $n + 1$ tautological 1-forms

$$\varphi^a_{(q,e)} = ds^2(e_a, q_*(\cdot)) \in \Omega^1(\mathcal{F}),$$

which form a basis for the semibasic 1-forms over Q. Next, there are globally defined, uniquely determined Levi-Civita connection forms $\varphi^a_b = -\varphi^b_a \in \Omega^1(\mathcal{F})$ satisfying

$$\begin{cases} d\varphi^a = -\varphi^a_b \wedge \varphi^b, \\ d\varphi^a_b = -\varphi^a_c \wedge \varphi^c_b + \frac{1}{2}R^a_{bcd}\varphi^c \wedge \varphi^d. \end{cases} \tag{4.14}$$

The functions R^a_{bcd} on $\mathcal{F}$ are the components of the Riemann curvature tensor with respect to different orthonormal frames, and satisfy

$$R^a_{bcd} + R^a_{bdc} = R^a_{bcd} + R^b_{acd} = R^a_{bcd} + R^a_{cdb} + R^a_{dbc} = 0.$$

We now distinguish the 1-form

$$\theta \overset{def}{=} \varphi^0 \in \Omega^1(\mathcal{F}),$$

which is the pullback via $\mathcal{F} \to M$ of the contact 1-form given the same name in (4.13). One of our structure equations now reads

$$d\theta = -\varphi^0_i \wedge \varphi^i. \tag{4.15}$$

This implies that the original $\theta \in \Omega^1(M)$ is actually a contact form, and also that $(\theta, \varphi^i, \varphi^0_i)$ is a basis for the semibasic 1-forms for $\mathcal{F} \to M$.

At this point, we can give the Poincaré-Cartan form for the prescribed mean curvature system. Namely, let $H \in C^\infty(Q)$ be a smooth function, and define on $\mathcal{F}$ the $(n + 1)$-form

$$\Pi \overset{def}{=} -\theta \wedge (\varphi^0_i \wedge \varphi_{(i)} - H\varphi).$$

Because H is the pullback of a function on Q, its derivative is of the form

$$dH = H_\nu \theta + H_i \varphi^i,$$

and using this and the structure equations (4.14), one can verify that Π is closed. Because Π is semibasic over M and closed, it is the pullback of a closed form on M, which is then a definite, neo-classical Poincaré-Cartan form. The associated Euler-Lagrange differential system then pulls back to $\mathcal{F}$ as

$$\mathcal{E}_H = \{\theta, d\theta, \varphi_i^0 \wedge \varphi_{(i)} - H\varphi\}.$$

While (4.15) shows that generic Legendre n-planes in M are defined by equations

$$\theta = 0, \quad \varphi_i^0 - h_{ij}\varphi^j = 0,$$

with $h_{ij} = h_{ji}$, integral n-planes in M for $\mathcal{E}_H$ are defined by the same equations, plus

$$h_{ii} = H.$$

The functions h_{ij} describing the tangent locus of a transverse Legendre submanifold of M are of course the coefficients of the second fundamental form of the corresponding submanifold of Q. Therefore, the transverse integral manifolds of $\mathcal{E}_H$ correspond locally to hypersurfaces in Q whose mean curvature h_{ii} equals the background function H. This will appear quite explicitly in what follows.

To investigate these integral manifolds, we employ the following apparatus. First consider the product

$$\mathcal{F} \times \mathbf{R}^{n(n+1)/2},$$

where $\mathbf{R}^{n(n+1)/2}$ has coordinates $h_{ij} = h_{ji}$, and inside this product define the locus

$$\mathcal{F}^{(1)} \stackrel{\text{def}}{=} \left\{ (f,h) \in \mathcal{F} \times \mathbf{R}^{n(n+1)/2} : h_{ii} = H \right\}.$$

To perform calculations, we want to extend our parallelization of $\mathcal{F}$ to $\mathcal{F}^{(1)}$. With a view toward reconstructing some of the bundle $B \to M$ associated to the Poincaré-Cartan form Π, we do this in a way that is as well-adapted to Π as possible.

On $\mathcal{F}^{(1)}$, we continue to work with θ, and define

$$\omega^i \stackrel{\text{def}}{=} \varphi^i,$$
$$\pi_i \stackrel{\text{def}}{=} \varphi_i^0 - h_{ij}\varphi^j.$$

With these definitions, we have

$$d\theta = -\pi_i \wedge \omega^i,$$
$$\Pi = -\theta \wedge \pi_i \wedge \omega_{(i)}.$$

Motivated by Riemannian geometry, we set

$$Dh_{ij} \stackrel{\text{def}}{=} dh_{ij} - h_{kj}\alpha_i^k - h_{ik}\alpha_j^k,$$

so that in particular $Dh_{ij} = Dh_{ji}$ and $Dh_{ii} = dH$. We also define the traceless part

$$Dh_{ij}^0 \stackrel{\text{def}}{=} Dh_{ij} - \tfrac{1}{n}\delta_{ij}dH.$$

Direct computations show that we will have exactly the structure equations (4.2, 4.3, 4.4) if we define

$$
\begin{aligned}
\alpha_j^i &= \varphi_j^i, \\
\rho &= -\tfrac{1}{2n} H\theta, \\
\delta_i &= -\tfrac{1}{2} H\pi_i + h_{ij}\pi_j + (h_{ik}h_{kj} - R_{ij0}^0 - \tfrac{1}{n}\delta_{ij}H_\nu)\omega^j, \\
\sigma_{ij} \wedge \omega^j &= (Dh_{ij}^0 + \tfrac{1}{n}\delta_{ij}H_k\omega^k + \tfrac{1}{2}R_{ijk}^0\omega^k) \wedge \omega^j,
\end{aligned}
$$

with $\sigma_{ij} = \sigma_{ji}$, $\sigma_{ii} = 0$. The last item requires some comment. Some linear algebra involving a Koszul complex shows that for any tensor V_{ijk} with $V_{ijk} = -V_{ikj}$ (this will be applied to $\tfrac{1}{2n}(\delta_{ij}H_k - \delta_{ik}H_j) + \tfrac{1}{2}R_{ijk}^0$), there is another tensor W_{ijk}, not unique, satisfying $W_{ijk} = W_{jik}$, $W_{iik} = 0$, and $\tfrac{1}{2}(W_{ijk} - W_{ikj}) = V_{ijk}$. This justifies the existence of σ_{ij} satisfying our requirements. The structure equations (4.2, 4.3, 4.4) resulting from these assignments have torsion coefficients

$$
T^{ijk} = 0, \quad U^{ij} = -\delta^{ij}, \quad S_j^i = -h_{ij} + \tfrac{1}{n}\delta_j^i H.
$$

The general calculations of §4.1.1 for the second variation can now be applied; note that we have the freedom to adapt coframes to a single integral submanifold in M of $\mathcal{E}_H$. Repeating those calculations verbatim leads to

$$
\frac{d^2}{dt^2}\bigg|_{t=0}\left(\int_{N_t}\Lambda\right) = -\int_{N_0} g(dg_i + n\rho g_i - g_j\alpha_i^j + g\delta_i) \wedge \omega_{(i)},
$$

where $F : N \times [0,1] \to M$ is a Legendre variation in M, F_0 is an integral manifold of $\mathcal{E}_H$, and the forms are all pullbacks of forms on $F^*(\mathcal{F}^{(1)})$ by a section of $F^*(\mathcal{F}^{(1)}) \to N \times [0,1]$, adapted along N_0 in the sense that

$$
\theta_{N_0} = g\,dt, \quad (\pi_i)_{N_0} = g_i dt.
$$

These imply that *restricted* to N_0, we have $\theta|_{N_0} = \pi_i|_{N_0} = 0$, and the preceding formula becomes

$$
\delta^2(\mathcal{F}_\Lambda)_{N_0}(g) = -\int_{N_0} g(dg_i - g_j\alpha_i^j + g(h_{ik}h_{kj} - \tfrac{1}{n}\delta_{ij}H_\nu - R_{ij0}^0)\omega^j) \wedge \omega_{(i)}.
$$

Recognizing that g can be thought of as a section of the normal bundle of the hypersurface $N \overset{F_0}{\hookrightarrow} M \overset{q}{\to} Q$, and that in this case g_i are the coefficients of its covariant derivative, this can be rewritten as

$$
\delta^2(\mathcal{F}_\Lambda)_{N_0}(g) = -\int_{N_0} (g\Delta g + g^2(||h||^2 - H_\nu - R_{ii0}^0))\omega, \tag{4.16}
$$

where Δ is the Riemannian Laplacian, and $||h||^2 = Tr(h^*h)$. Here, the extrinsic curvature function K appears as the quantity $||h||^2 - H_\nu - R_{ii0}^0$. Notice that we have actually calculated this second variation without ever determining the functional Λ. In case the ambient manifold Q is flat Euclidean space, if the background function H is a constant and the variation g is compactly supported in the interior of N, this simplifies to

$$\left.\frac{d^2}{dt^2}\right|_{t=0} \left(\int_{N_t} \Lambda\right) = -\int_{N_0} (g\Delta g + g^2||h||^2)\omega$$

$$= \int_{N_0} (||\nabla g||^2 - g^2||h||^2)\omega.$$

Even for the minimal surface equation $H = 0$, we cannot conclude from this formula alone that a solution locally minimizes area.

4.1.5 Conditions for a Local Minimum

We now discuss some conditions under which an integral manifold $N \hookrightarrow M$ of an Euler-Lagrange system $\mathcal{E}_\Lambda \subset \Omega^*(M)$ is a local minimum for the functional $\mathcal{F}_\Lambda$, in the sense that $\mathcal{F}_\Lambda(N) < \mathcal{F}_\Lambda(N')$ for all Legendre submanifolds N' near N. However, there are two natural meanings for "near" in this context, and this will yield two notions of local minimum. Namely, we will say that $\mathcal{F}_\Lambda$ has a *strong* local minimum at N if the preceding inequality holds whenever N' is C^0-close to N, while $\mathcal{F}_\Lambda$ has a *weak* local minimum at N if the preceding inequality holds only among the narrower class of N' which are C^1-close to N.[2]

Our goal is to illustrate how the Poincaré-Cartan form may be used to understand in a simple geometric manner some classical conditions on extrema. Specifically, we will introduce the notion of a *calibration* for an integral manifold of the Euler-Lagrange system; its existence (under mild topological hypotheses) implies that the integral manifold is a strong local minimum. Under certain classical conditions for a local minimum, we will use the Poincaré-Cartan form to construct an analogous *weak calibration*. Finally, our geometric description of the second variation formula highlights the Jacobi operator $Jg = -\Delta_c g + Kg\omega$, and some linear analysis shows that the positivity of the first eigenvalue of J implies the classical conditions.

Let Π be a neo-classical Poincaré-Cartan form Π on a contact manifold (M, I). There is a local foliation $M \to Q$, and we can choose coordinates to have $(x^i, z) \in Q \subset \mathbf{R}^n \times \mathbf{R}$, $(x^i, z, p_i) \in M \subset J^1(\mathbf{R}^n, \mathbf{R})$, $\theta = dz - p_i dx^i \in \Gamma(I)$, and a Lagrangian potential

$$\Lambda = L(x, z, p)dx + \theta \wedge L_{p_i} dx_{(i)} \in \Omega^n(M)$$

whose Poincaré-Cartan form is

$$\Pi = d\Lambda = \theta \wedge (-dL_{p_i} \wedge dx_{(i)} + L_z dx).$$

---

[2] A thorough, coordinate-based discussion of the relevant analysis can be found in [GH96].

We will confine our discussion to a domain where this classical description holds. We may regard an integral manifold of the Euler-Lagrange system $\mathcal{E}_\Lambda$ as a submanifold $N_0 \hookrightarrow Q$ given by the graph $\{(x, z_0(x)) : x \in U\}$ of a solution to the Euler-Lagrange equation over some open $U \subset \mathbf{R}^n$. It has a natural 1-jet extension $N_0^{(1)} \hookrightarrow M$, equal to $\{(x, z_0(x), \nabla z_0(x)) : x \in U\}$, which is an integral manifold of $\mathcal{E}_\Lambda$ in the sense discussed previously. We define a *strong neighborhood* of N_0 to be the collection of hypersurfaces in Q lying in some open neighborhood of N_0 in Q, and a *weak neighborhood* of N_0 to be the collection of hypersurfaces $N \hookrightarrow Q$ whose 1-jet prolongations $N^{(1)}$ lie in some open neighborhood of $N_0^{(1)}$ in M. Whether or not a given stationary submanifold N_0 is minimal depends on which of these two classes of competing submanifolds one studies.

Starting with strong neighborhoods, we fix a neighborhood $W \subset Q$ of a stationary submanifold $N_0 \subset Q$ for Λ, and introduce the following useful notion.

Definition 4.1 *A* calibration *for* (Λ, N_0) *is an n-form* $\tilde{\Lambda} \in \Omega^n(W)$ *satisfying*

- $d\tilde{\Lambda} = 0$;

- $\tilde{\Lambda}|_{N_0} = \Lambda|_{N_0^{(1)}}$;

- $\tilde{\Lambda}|_E \leq \Lambda_E$ *for each n-plane $E^n \subset T_q W$.*

In the right-hand side of the last inequality, we are regarding the n-plane E as specifying a point of $M \subset G_n(TQ)$ over $q \in Q$, and evaluating $\Lambda_{(q,E)}$ on any tangent n-plane $E' \subset T_{(q,E)}M$ projecting one-to-one into $T_q Q$; the value is independent of the choice of E', because Λ is semibasic over Q. In particular, the third condition says that the integral of $\tilde{\Lambda}$ on any $N \subset W \subset Q$ will not exceed the integral of Λ on $N^{(1)} \subset M$.

In both the strong and weak settings, we will only have N_0 compete against submanifolds having the same boundary. For this reason, we assume that $W \supset N_0$ is chosen so that $\partial N_0 = N_0 \cap \partial W$, and that $(N_0, \partial N_0)$ generates the relative homology $H_n(W, \partial W; \mathbf{Z})$.

Proposition 4.1 *If there exists a calibration $\tilde{\Lambda}$ for (Λ, N_0), then $\mathcal{F}_\Lambda(N_0) \leq \mathcal{F}_\Lambda(N)$ for every hypersurface $N \hookrightarrow W \subset Q$ satisfying $\partial N = \partial N_0$.*

We then say that N_0 is a *strong (but not strict!) local minimum* for $\mathcal{F}_\Lambda$.

Proof. We simply calculate

$$
\begin{aligned}
\mathcal{F}_\Lambda(N_0) &= \int_{N_0^{(1)}} \Lambda \\
&= \int_{N_0} \tilde{\Lambda} \\
&= \int_N \tilde{\Lambda} \\
&\leq \int_{N^{(1)}} \Lambda \\
&= \mathcal{F}_\Lambda(N).
\end{aligned}
$$

The third equality uses Stokes' theorem, which applies because our topological hypothesis on W implies that the cycle $N - N_0$ in W is a boundary. □

The question of when one can find a calibration naturally arises. For this, we use the following classical concept.

Definition 4.2 *A field for* (Λ, N_0) *is a neighborhood* $W \subset Q$ *of* N_0 *with a smooth foliation by a 1-parameter family* $F : N \times (-\varepsilon, \varepsilon) \to W$ *of integral manifolds of* $\mathcal{E}_\Lambda$.

This family does *not* have a fixed boundary. We retain the topological hypotheses on W used in Proposition 4.1, and have the following.

Proposition 4.2 *If there exists a field for* (Λ, N_0)*, then there exists a closed form* $\tilde{\Lambda} \in \Omega^n(W)$ *such that* $\tilde{\Lambda}|_{N_0} = \Lambda|_{N_0^{(1)}}$.

$\tilde{\Lambda}$ is then a calibration if it additionally satisfies the third condition, $\tilde{\Lambda}|_E \leq \Lambda_E$. In the proof, we will explicitly construct $\tilde{\Lambda}$ using the Poincaré-Cartan form.

Proof. The field $F : N \times (-\varepsilon, \varepsilon) \to Q$ may be thought of as a family of graphs

$$N_t = \{(x, z(x, t))\},$$

where each $z(\cdot, t)$ is a solution of the Euler-Lagrange equations, and the domain of $z(\cdot, t)$ may depend on $t \in (-\varepsilon, \varepsilon)$. Because each point of W lies on exactly one of these graphs, we can define a 1-jet lift $F' : W \to M$, given by

$$(x, z) = (x, z(x, t)) \mapsto (x, z(x, t), \nabla_x z(x, t)).$$

Let $\tilde{\Lambda} = (F')^* \Lambda \in \Omega^n(W)$. Then it is clear that $\tilde{\Lambda}|_{N_0} = \Lambda|_{N_0^{(1)}}$, and to show that $\tilde{\Lambda}$ is closed, we need to see that $(F')^* \Pi = d\tilde{\Lambda} = 0$. This holds because $\Pi = \theta \wedge \Psi$ is quadratic in an ideal of forms vanishing on each leaf $F_t^{(1)} : N \to M$; more concretely, each of $(F')^* \theta$ and $(F')^* \Psi$ must be a multiple of dt, so their product vanishes. □

General conditions for $\tilde\Lambda = (F')^*\Lambda$ to be a calibration, and for N_0 to therefore be a strong local minimum, are not clear. However, we can still use the preceding to detect weak local minima.

Proposition 4.3 *Under the hypotheses of Propositions 4.1 and 4.2, if*

$$L_{p_i p_j}(x, z_0(x), \nabla z_0(x))\xi_i \xi_j \geq c\|\xi\|^2, \qquad (4.17)$$

for some constant $c > 0$ and all (ξ_i), then $\tilde\Lambda|_E \leq \Lambda_E$ for all $E \subset T_q Q$ sufficiently near $T_q N_0$, with equality if and only if $E = T_q N_0$. Furthermore, $\mathcal{F}_\Lambda(N_0) < \mathcal{F}_\Lambda(N)$ for all $N \neq N_0$ in a weak neighborhood of N_0.

The first statement allows us to think of $\tilde\Lambda$ as a *weak calibration* for (Λ, N_0). The proof of the second statement from the first will use Stokes' theorem in exactly the manner of Proposition 4.1.

Proof. The positivity of the ∇z-Hessian of L suggests that we define the *Weierstrass excess function*

$$E(x, z, p, q) \stackrel{def}{=} L(x, z, p) - L(x, z, q) - \sum(p_i - q_i)L_{q_i}(x, z, q),$$

which is the second-order remainder in a Taylor series expansion for L. This function will appear in a more detailed expression for $\pi^*\tilde\Lambda = (F' \circ \pi)^*\Lambda \in \Omega^n(M)$, computed modulo $\{I\}$. We write

$$F'(x^i, z) = (x^i, z, q_i(x, z)) \in M,$$

where (x^i, z, p_i) are the usual coordinates on M, and the functions $q_i(x, z)$ are the partial derivatives of the field elements $z(x, t)$. We have

$$
\begin{aligned}
\pi^*\tilde\Lambda &= \pi^* \circ F'^*(L\,dx + \theta \wedge L_{p_i}\,dx_{(i)}) \\
&= L(x^i, z, q_i(x, z))dx + (dz - q_i(x, z)dx^i) \wedge L_{p_i}(x^i, z, q_i(x, z))dx_{(i)} \\
&\equiv \left(L(x^i, z, q_i(x, z)) + (p_i - q_i(x, z))L_{p_i}(x^i, z, q_i(x, z))\right) dx \pmod{\{I\}} \\
&\equiv \Lambda - E(x^i, z, p_i, q_i(x, z))dx \pmod{\{I\}}.
\end{aligned}
$$

The hypothesis (4.17) on the Hessian $F_{p_i p_j}$ implies that for each (x^i, z), and p_i sufficiently close to $q_i(x, z)$, the second-order remainder satisfies

$$E(x^i, z, p_i, q_i(x, z)) \geq 0,$$

with equality if and only if $p_i = q_i(x, z)$. The congruence of $\pi^*\tilde\Lambda$ and $\Lambda - E\,dx$ modulo $\{I\}$ then implies our first statement.

For the second statement, we use the Stokes' theorem argument:

$$
\begin{aligned}
\int_{N_0^{(1)}} \Lambda &= \int_{N_0} \tilde{\Lambda} \\
&= \int_N \tilde{\Lambda} \\
&= \int_{N^{(1)}} (\Lambda - E(x^i, z, p_i, q_i(x, z)) dx) \\
&\leq \int_{N^{(1)}} \Lambda,
\end{aligned}
$$

with equality in the last step if and only if $N_0 = N$. $\square$

This proof shows additionally that if the Weierstrass excess function satisfies $E(x^i, z, p_i, q_i) > 0$ for all $p \neq q$, then N_0 is a strong (and strict) local minimum for $\mathcal{F}_\Lambda$.

So far, we have shown that if we can cover some neighborhood of a stationary submanifold N_0 with a field, then we can construct an n-form $\tilde{\Lambda}$, whose calibration properties imply extremal properties of N_0. It is therefore natural to ask when such a field exists, and the answer to this involves some analysis of the *Jacobi operator*. We will describe the operator, and hint at the analysis.

The Jacobi operator acts on sections of a density line bundle on a given integral manifold N_0 of the Euler-Lagrange system, with its induced conformal structure. Specifically,

$$
J : D^{\frac{n-2}{2n}} \to D^{\frac{n+2}{2n}}
$$

is the differential operator given by

$$
Jg = -\Delta_c g - gK,
$$

where Δ_c is the conformal Laplacian, and K is the curvature invariant introduced in §4.1.2. The second variation formula (4.8) then reads

$$
\delta^2(\mathcal{F}_\Lambda)_{N_0}(g) = \int_{N_0} g\, Jg\, \omega.
$$

The main geometric fact is:

> *The Jacobi operator gives the (linear) variational equations for integral manifolds of the Euler-Lagrange system $\mathcal{E}_\Lambda$.*

This means the following. Let $F : N \times [0,1] \to M$ be a Legendre variation of the Λ-stationary submanifold F_0—not necessarily having fixed boundary—and let $g = \left(\frac{\partial F}{\partial t}\big|_{t=0}\right) \lrcorner\, \theta$, as usual. Then our previous calculations imply that $Jg = 0$ if and only if

$$
\mathcal{L}_{\frac{\partial}{\partial t}}(F^*\Psi)|_{t=0} = 0.
$$

We might express this condition by saying that F_t is an integral manifold for $\mathcal{E}_\Lambda = \mathcal{I} + \{\Psi\}$ modulo $O(t^2)$.

We now indicate how a condition on the Jacobi operator of N_0 can imply the existence of a field near N_0. Consider the eigenvalue problem

$$Jg = -\Delta_c g - gK = \lambda g, \qquad g \in C_0^\infty(N),$$

for smooth, fixed boundary variations. It is well-known that J has a discrete spectrum bounded from below, $\lambda_1 < \lambda_2 < \cdots$, with $\lambda_k \to \infty$ and with finite-dimensional eigenspaces. We consider the consequences of the assumption

$$\lambda_1 > 0.$$

Because $\lambda_1 = \inf\{\int_N g\, Jg\, \omega : ||g||_{L^2} = 1\}$, the assumption $\lambda_1 > 0$ is equivalent to

$$\delta^2(\mathcal{F}_\Lambda)_N(g) > 0, \qquad \text{for } g \neq 0.$$

The main analytic result is the following.

Proposition 4.4 *If $\lambda_1 > 0$, then given $g_0 \in C^\infty(\partial N)$, there is a unique solution $g \in C^\infty(N)$ to the boundary value problem*

$$Jg = 0, \qquad g|_{\partial N} = g_0.$$

Furthermore, if $g_0 > 0$ on ∂N, then this solution satisfies $g > 0$ on N.

The existence and uniqueness statements follow from standard elliptic theory. The point is that we can compare the second variation $\delta^2(\mathcal{F}_\Lambda)_N(g)$ to the Sobolev norm $||g||_1^2 = \int_N(||\nabla g||^2 + |g|^2)\omega$, and if $\lambda_1 > 0$, then there are constants $c_1, c_2 > 0$ such that

$$c_1||g||_1^2 \leq \int_N g\, Jg\, \omega \leq c_2||g||_1^2.$$

The Schauder theory gives existence and uniqueness in this situation.

Less standard is the positivity of the solution g under the assumption that $g|_{\partial N} > 0$, and this is crucial for the existence of a field. Namely, a further implicit function argument using elliptic theory guarantees that the variation g is tangent to an arc of integral manifolds of $\mathcal{E}$, and the fact that $g \neq 0$ implies that near the initial N, this arc defines a field. For the proof of the positivity of g, and details of all of the analysis, see [GH96].

4.2   Euler-Lagrange PDE Systems

Up to this point, we have studied geometric aspects of first-order Lagrangian functionals

$$\mathcal{F}_L(z) = \int_\Omega L\left(x^i, z, \tfrac{\partial z}{\partial x^i}\right) dx, \qquad \Omega \subset \mathbf{R}^n, \tag{4.18}$$

where $x = (x^1, \ldots, x^n)$ and $z = z(x)$ is a scalar function. In this section, we consider the more general situation of functionals

$$\mathcal{F}_L(z) = \int_\Omega L\left(x^i, z^\alpha(x), \tfrac{\partial z^\alpha}{\partial x^i}(x)\right) dx, \qquad \Omega \subset \mathbf{R}^n, \tag{4.19}$$

where now $z(x) = (z^1(x), \ldots, z^s(x))$ is an $\mathbf{R}^s$-valued function of $x = (x^i)$, and $L = L(x^i, z^\alpha, p_i^\alpha)$ is a smooth function on $\mathbf{R}^{n+s+ns}$. The Euler-Lagrange equations describing maps $z : \Omega \to \mathbf{R}^s$ which are stationary for $\mathcal{F}_L$ under all fixed-boundary variations form a PDE system

$$\frac{\partial L}{\partial z^\alpha} - \sum_i \frac{d}{dx^i}\left(\frac{\partial L}{\partial p_i^\alpha}\right) = 0, \qquad \alpha = 1, \ldots, s. \tag{4.20}$$

In the scalar case $s = 1$, we have examined the geometry of the equivalence class of $\mathcal{F}_L$ under contact transformations and found the canonically defined Poincaré-Cartan form to be of considerable use. In this section, we describe a generalization of the Poincaré-Cartan form for $s \geq 1$. Geometrically, we study functionals on the space of compact submanifolds of codimension s, in an $(n+s)$-dimensional manifold with local coordinates (x^i, z^α).

An immediate difference between the cases $s = 1$ and $s \geq 2$ is that in the latter case, there are no proper contact transformations of $\mathbf{R}^{n+s+ns}$; that is, the only smooth maps $x' = x'(x, z, p)$, $z' = z'(x, z, p)$, $p' = p'(x, z, p)$ for which

$$\{dz^\alpha - \sum p_i^\alpha dx^i\} = \{dz^{\alpha'} - \sum p_i^{\alpha'} dx^{i'}\}$$

are point transformations $x' = x'(x, z)$, $z' = z'(x, z)$, with $p' = p'(x, z, p)$ determined by the chain rule. We will explain why this is so, and later, we will see that in case $s = 1$ our original contact-invariant Poincaré-Cartan form still appears naturally in the more limited context of point transformations. Our first task, however, is to introduce the geometric setting for studying functionals (4.19) subject to point transformations, analogous to our use of contact manifolds for (4.18).

Throughout this section, we have as always $n \geq 2$ and we use the index ranges $1 \leq i, j \leq n$, $1 \leq \alpha, \beta \leq s$.

4.2.1 Multi-contact Geometry

Having decided to apply point transformations to the functional (4.19), we interpret $z(x) = (z^\alpha(x^i))$ as corresponding to an n-dimensional submanifold of $\mathbf{R}^{n+s}$. The first derivatives $p_i^\alpha = \frac{\partial z^\alpha}{\partial x^i}$ specify the tangent n-planes of this submanifold. This suggests our first level of geometric generalization.

Let X be a manifold of dimension $n + s$, and let $G_n(TX) \xrightarrow{\pi} X$ be the Grassmannian bundle of n-dimensional subspaces of tangent spaces of X; that is, a point of $G_n(TX)$ is of the form

$$m = (p, E), \qquad p \in X, \ E^n \subset T_pX.$$

Any diffeomorphism of X induces a diffeomorphism of $G_n(TX)$, and either of these diffeomorphisms will be called a *point transformation*.

We can define on $G_n(TX)$ two Pfaffian systems $I \subset J \subset T^*(G_n(TX))$, of ranks s and $n + s$, respectively, which are canonical in the sense that they are preserved by any point transformation. First, $J = \pi^*(T^*X)$ consists of all forms that are semibasic over X; J is integrable, and its maximal integral submanifolds are the fibers of $G_n(TX) \to X$. Second, we define I at a point $(p, E) \in G_n(TX)$ to be

$$I_{(p,E)} = \pi_p^*(E^\perp),$$

where $E^\perp \subset T_p^* X$ is the s-dimensional annihilator of the subspace $E \subset T_p X$. I is not integrable, and to understand its integral submanifolds, note that any n-dimensional immersion $\iota : N \hookrightarrow X$ has a 1-jet lift $\iota^{(1)} : N \hookrightarrow G_n(TX)$. In fact, such lifts are the transverse integral submanifolds of the Pfaffian system $I \subset T^*(G_n(TX))$.

To see this explicitly, choose local coordinates (x^i, z^α) on $U \subset X^{n+s}$. These induce local coordinates $(x^i, z^\alpha, p_i^\alpha)$ corresponding to the n-plane $E \subset T_{(x^i, z^\alpha)} U$ defined as

$$E = \{dz^1 - p_i^1 dx^i, \ldots, dz^s - p_i^s dx^i\}^\perp.$$

These coordinates are defined on a dense open subset of $\pi^{-1}(U) \subset G_n(TX)$, consisting of n-planes $E \subset TX$ for which $dx^1 \wedge \cdots \wedge dx^n|_E \neq 0$. In terms of these local coordinates on $G_n(TX)$, our Pfaffian systems are

$$
\begin{aligned}
J &= \{dx^i, dz^\alpha\}, \\
I &= \{dz^\alpha - p_i^\alpha dx^i\}.
\end{aligned}
$$

An immersed submanifold $N^n \hookrightarrow U$ for which $dx^1 \wedge \cdots \wedge dx^n|_N \neq 0$ may be regarded as a graph

$$N = \{(x^i, z^\alpha) : z^\alpha = f^\alpha(x^1, \ldots, x^n)\}.$$

Its lift to $N \hookrightarrow G_n(TX)$ lies in the domain of the coordinates $(x^i, z^\alpha, p_i^\alpha)$, and equals the 1-jet graph

$$N^{(1)} = \{(x^i, z^\alpha, p_i^\alpha) : z^\alpha = f^\alpha(x^1, \ldots, x^n), \ p_i^\alpha = \tfrac{\partial f^\alpha}{\partial x^i}(x^1, \ldots, x^n)\}. \quad (4.21)$$

Clearly this lift is an integral submanifold of I. Conversely, a submanifold $N^{(1)} \hookrightarrow \pi^{-1}(U) \subset G_n(TX)$ on which $dz^\alpha - \sum p_i^\alpha dx^i = 0$ and $dx^1 \wedge \cdots \wedge dx^n \neq 0$ is necessarily given locally by a graph of the form (4.21). The manifold $M = G_n(TX)$ with its Pfaffian systems $I \subset J$ is our standard example of a *multi-contact manifold*. This notion will be defined shortly, in terms of the following structural properties of the Pfaffian systems.

Consider on $G_n(TX)$ the differential ideal $\mathcal{I} = \{I, dI\} \subset \Omega^*(G_n(TX))$ generated by $I \subset T^*(G_n(TX))$. If we set

$$\bar{\theta}^\alpha = dz^\alpha - p_i^\alpha dx^i, \quad \bar{\omega}^i = dx^i, \quad \bar{\pi}_i^\alpha = dp_i^\alpha,$$

then we have the structure equations

$$d\bar{\theta}^\alpha \equiv -\bar{\pi}_i^\alpha \wedge \bar{\omega}^i \pmod{\{I\}}, \quad 1 \leq \alpha \leq s. \quad (4.22)$$

It is not difficult to verify that the set of all coframings $(\theta^\alpha, \omega^i, \pi_i^\alpha)$ on $G_n(TX)$ for which

- $\theta^1, \ldots, \theta^s$ generate I,

- $\theta^1, \ldots, \theta^s, \omega^1, \ldots, \omega^n$ generate J, and

- $d\theta^\alpha \equiv -\pi_i^\alpha \wedge \omega^i \pmod{\{I\}}$

are the local sections of a G-structure on $G_n(TX)$. Here $G \subset GL(n+s+ns, \mathbf{R})$ may be represented as acting on $(\theta^\alpha, \omega^i, \pi_i^\alpha)$ by

$$\begin{cases} \bar{\theta}^\alpha = a_\beta^\alpha \theta^\beta, \\ \bar{\omega}^i = c_\beta^i \theta^\beta + b_j^i \omega^j, \\ \bar{\pi}_i^\alpha = d_{i\beta}^\alpha \theta^\beta + e_{kj}^\alpha (b^{-1})_i^k \omega^j + a_\beta^\alpha \pi_j^\beta (b^{-1})_i^j, \end{cases} \tag{4.23}$$

where $(a_\beta^\alpha) \in GL(s, \mathbf{R})$, $(b_j^i) \in GL(n, \mathbf{R})$, and $e_{ij}^\alpha = e_{ji}^\alpha$. From these properties we make our definition.

Definition 4.3 *A* multi-contact manifold *is a manifold* M^{n+s+ns}, *with a G-structure as in (4.23), whose sections* $(\theta^\alpha, \omega^i, \pi_i^\alpha)$ *satisfy*

$$\begin{align} d\theta^\alpha &\equiv -\pi_i^\alpha \wedge \omega^i \pmod{\{\theta^1, \ldots, \theta^s\}}, \tag{4.24} \\ d\omega^i &\equiv 0 \pmod{\{\theta^1, \ldots, \theta^s, \omega^1, \ldots, \omega^n\}}. \tag{4.25} \end{align}$$

Note that the G-structure determines Pfaffian systems $I = \{\theta^1, \ldots, \theta^s\}$ and $J = \{\theta^1, \ldots, \theta^s, \omega^1, \ldots, \omega^n\}$, and we may often refer to (M, I, J) as a multi-contact manifold, implicitly assuming that J is integrable and that there are coframings for which the structure equations (4.22) hold. The integrability of J implies that locally in M one can define a smooth leaf space X^{n+s} and a surjective submersion $M \to X$ whose fibers are integral manifolds of J. When working locally in a multi-contact manifold, we will often make reference to this quotient X.

It is not difficult to show that any multi-contact structure (M, I, J) is locally equivalent to that of $G_n(TX)$ for a manifold X^{n+s}. The integrability of J implies that there are local coordinates $(x^i, z^\alpha, q_i^\alpha)$ for which dx^i, dz^α generate J. We can relabel the x^i, z^α to assume that $dz^\alpha - p_i^\alpha dx^i$ generate I for some functions $p_i^\alpha(x, z, q)$. The structure equations then imply that $dx^i, dz^\alpha, dp_i^\alpha$ are linearly independent, so on a possibly smaller neighborhood in M, we can replace the coordinates q_i^α by p_i^α, and this exhibits our structure as equivalent to that of $G_n(TX)$.

We will see below that if $s \geq 2$, then the Pfaffian system I of a multi-contact manifold uniquely determines the larger system J. Also, if $s \geq 3$, the hypothesis (4.25) that $J = \{\theta^\alpha, \omega^i\}$ is integrable is not necessary; it is easily seen to be a consequence of the structure equation (4.24). However, in the case $s = 2$, J is determined by I but is not necessarily integrable; our study of Euler-Lagrange systems will not involve this exceptional situation, so we have ruled it out in our definition.

It is not at all obvious how one can determine, given a Pfaffian system I of rank s on a manifold M of dimension $n + s + ns$, whether I comes from a multi-contact structure; deciding whether structure equations (4.24) can be satisfied for some generators of I is a difficult problem. Bryant has given easily evaluated intrinsic criteria characterizing such I, generalizing the Pfaff theorem's normal form for contact manifolds, but we shall not need this here (see Ch. II, §4 of [B+91]).

Aside from those of the form $G_n(TX)$, there are two other kinds of multi-contact manifolds in common use. One is $J^1(Y^n, Z^s)$, the space of 1-jets of maps from an n-manifold Y to an s-manifold Z. The other is $J^1_\Gamma(E^{n+s}, Y^n)$, the space of 1-jets of sections of a fiber bundle $E \to Y$ with base of dimension n and fiber Z of dimension s. These are distinguished by the kinds of coordinate changes considered admissible in each case; to the space $J^1(Y^n, Z^s)$, one would apply prolonged classical transformations $x'(x)$, $z'(z)$, while to $J^1_\Gamma(E^{n+s}, Y^n)$, one would apply prolonged gauge transformations $x'(x)$, $z'(x, z)$. These are both smaller classes than the point transformations $x'(x, z)$, $z'(x, z)$ that we apply to $G_n(TX)$, the space of 1-jets of n-submanifolds in X^{n+s}.

Recall our claim that in the multi-contact case $s \geq 2$, every contact transformation is a prolonged point transformation. This is the same as saying that any local diffeomorphism of M which preserves the Pfaffian system I also preserves J; for a local diffeomorphism of M preserving J must induce a diffeomorphism of the local quotient space X, which in turn uniquely determines the original local diffeomorphism of M. To see why a local diffeomorphism preserving I must preserve J, we will give an intrinsic construction of J in terms of I alone, for the local model $G_n(TX)$. First, define for any 2-form $\Psi \in \bigwedge^2(T^*_m M)$ the space of 1-forms

$$\mathcal{C}(\Psi) = \{V \lrcorner \Psi : V \in T_m M\}.$$

This is a pointwise construction. We apply it to each element of the vector space

$$\{\lambda_\alpha d\theta^\alpha : (\lambda^\alpha) \in \mathbf{R}^s\},$$

intrinsically given as the quotient of $\mathcal{I}_2$, the degree-2 part of the multi-contact differential ideal, by the subspace $\{I\}_2$, the degree-2 part of the *algebraic* ideal $\{I\}$. For example,

$$\mathcal{C}(\overline{d\theta^\alpha}) \equiv \mathrm{Span}\{\pi_i^\alpha, \omega^i\} \pmod{\{I\}}.$$

The intersection

$$\bigcap_{\Theta \in \mathcal{I}_2/\{I\}_2} \mathcal{C}(\Theta)$$

is a well-defined subbundle of T^*M/I. If $s \geq 2$, then its preimage in T^*M is $J = \{\bar\theta^\alpha, \bar\omega^i\}$, as is easily seen using the structure equations (4.22). Any local diffeomorphism of M preserving I therefore preserves $\mathcal{I}_2$, $\mathcal{I}_2/\{I\}_2$, $\bigcap \mathcal{C}(\Theta)$, and finally J, which is what we wanted to prove. Note that in the contact case $s = 1$, $\bigcap \mathcal{C}(\Theta) \equiv \mathcal{C}(d\theta) \equiv \{\pi_i, \omega^i\}$ modulo $\{I\}$, so instead of this construction giving J, it gives all of T^*M. In this case, introducing J in the definition of a multi-contact manifold reduces our pseudogroup from contact transformations to point transformations.

We have given a generalization of the notion of a contact manifold to accomodate the study of submanifolds of codimension greater than one. There is a further generalization to higher-order contact geometry, which is the correct setting for studying higher-order Lagrangian functionals, and we will consider it briefly in the next section.

In what follows, we will carry out the discussion of functionals modeled on (4.19) on a general multi-contact manifold (M, I, J), but the reader can concentrate on the case $M = G_n(TX)$.

4.2.2   Functionals on Submanifolds of Higher Codimension

Returning to our functional (4.19), we think of the integrand $L(x^i, z^\alpha, p^i_\alpha)dx$ as an n-form on a dense open subset of the multi-contact manifold $G_n(T\mathbf{R}^{n+s})$. Note that this n-form is semibasic for the projection $G_n(T\mathbf{R}^{n+s}) \to \mathbf{R}^{n+s}$, and that any n-form congruent to $L(x^i, z^\alpha, p^i_\alpha)dx$ modulo $\{dz^\alpha - p^\alpha_i dx^i\}$ gives the same classical functional. This suggests the following.

Definition 4.4 *A* Lagrangian *on a multi-contact manifold* (M, I, J) *is a smooth section* $\Lambda \in \Gamma(M, \bigwedge^n J) \subset \Omega^n(M)$. *Two Lagrangians are* equivalent *if they are congruent modulo* $\{I\}$.

An equivalence class $[\Lambda]$ of Lagrangians corresponds to a section of the vector bundle $\bigwedge^n(J/I)$. It also defines a functional on the space of compact integral manifolds (possibly with boundary) of the Pfaffian system I by

$$\mathcal{F}_\Lambda(N) = \int_N \Lambda,$$

where Λ is any representative of the class. The notion of *divergence equivalence* of Lagrangians will appear later. In the discussion in Chapter 1 of the scalar case $s = 1$, we combined these two types of equivalence by emphasizing a characteristic cohomology class in $H^n(\Omega^*(M)/\mathcal{I})$, and used facts about symplectic linear algebra to investigate these classes. However, the analogous "multi-symplectic" linear algebra that is appropriate for the study of multi-contact geometry is still poorly understood.[3]

Our goal is to associate to any functional $[\Lambda] \in \Gamma(M, \bigwedge^n(J/I))$ a Lagrangian $\Lambda \in \Gamma(M, \bigwedge^n J) \subset \Omega^n(M)$, not necessarily uniquely determined, whose exterior derivative $\Pi = d\Lambda$ has certain favorable properties and *is* uniquely determined by $[\Lambda]$. Among these properties are:

- $\Pi \equiv 0 \pmod{\{I\}}$;

- Π is preserved under any diffeomorphism of M preserving I, J, and $[\Lambda]$;

---

[3]Results in [Gra00] may illuminate this issue, along with some others that will come up in the following discussion.

- Π depends only on the divergence-equivalence class of $[\Lambda]$;

- $\Pi = 0$ if and only if the Euler-Lagrange equations for $[\Lambda]$ are trivial.

Triviality of the Euler-Lagrange equations means that every compact integral manifold of $\mathcal{I} \subset \Omega^*(M)$ is stationary for $\mathcal{F}_\Lambda$ under fixed-boundary variations. Some less obvious ways in which such Π could be useful are the following, based on our experience in the scalar case $s = 1$:

- in Noether's theorem, where one would hope for $v \mapsto v \lrcorner \Pi$ to give an isomorphism from a Lie algebra of symmetries to a space of conservation laws;

- in the inverse problem, where one can try to detect equations that are locally of Euler-Lagrange type not by finding a Lagrangian, but by finding a Poincaré-Cartan form inducing the equations;

- in the study of local minimization, where it could help one obtain a calibration in terms of a field of stationary submanifolds.

Recall that in the case of a contact manifold, we replaced any Lagrangian $\Lambda \in \Omega^n(M)$ by
$$\Lambda - \theta \wedge \beta,$$
the unique form congruent to Λ (mod $\{I\}$) with the property that
$$d\Lambda \equiv 0 \quad (\text{mod } \{I\}).$$

What happens in the multi-contact case? Any Lagrangian $\Lambda_0 \in \Gamma(M, \bigwedge^n J)$ is congruent modulo $\{I\}$ to a form (in local coordinates)
$$L(x^i, z^\alpha, p_i^\alpha)dx,$$
and motivated by the scalar case, we consider the equivalent form
$$\Lambda = L\,dx + \theta^\alpha \wedge \tfrac{\partial L}{\partial p_i^\alpha}dx_{(i)}, \tag{4.26}$$
which has exterior derivative
$$d\Lambda = \theta^\alpha \wedge \left(\tfrac{\partial L}{\partial z^\alpha}dx - d\left(\tfrac{\partial L}{\partial p_i^\alpha} \right) \wedge dx_{(i)} \right). \tag{4.27}$$

This suggests the following definition.

Definition 4.5 *An* admissible lifting *of a functional* $[\Lambda] \in \Gamma(M, \bigwedge^n(J/I))$ *is a Lagrangian* $\Lambda \in \Gamma(M, \bigwedge^n J)$ *representing the class* $[\Lambda]$ *and satisfying* $d\Lambda \in \{I\}$.

The preceding calculation shows that locally, every functional $[\Lambda]$ has an admissible lifting. Unfortunately, the admissible lifting is generally not unique. This will be addressed below, but first we show that any admissible lifting is adequate for calculating the first variation and the Euler-Lagrange system of the functional $\mathcal{F}_\Lambda$.

We mimic the derivation in Chapter 1 of the Euler-Lagrange differential system in the scalar case $s = 1$. Suppose that we have a 1-parameter family $\{N_t\}$ of integral manifolds of the multi-contact Pfaffian system I, given as a smooth map

$$F : N \times [0,1] \to M,$$

for which each $F_t = F|_{N \times \{t\}} : N \hookrightarrow M$ is an integral manifold of I and such that $F|_{\partial N \times [0,1]}$ is independent of t. Then choosing generators $\theta^\alpha \in \Gamma(I)$, $1 \le \alpha \le s$, we have

$$F^* \theta^\alpha = G^\alpha dt$$

for some functions G^α on $N \times [0,1]$. As in the contact case, it is not difficult to show that any collection of functions g^α supported in the interior of N can be realized as $G^\alpha|_{t=0}$ for some 1-parameter family N_t.

The hypothesis that Λ is an admissible lifting means that we can write

$$d\Lambda = \sum \theta^\alpha \wedge \Psi_\alpha$$

for some $\Psi_\alpha \in \Omega^n(M)$. Then we can proceed as in §1.2 to calculate

$$\frac{d}{dt}\bigg|_{t=0} \left(\int_{N_t} F_t^* \Lambda \right) = \int_{N_0} \mathcal{L}_{\frac{\partial}{\partial t}}(F^* \Lambda)$$
$$= \int_{N_0} \frac{\partial}{\partial t} \lrcorner (\theta^\alpha \wedge \Psi_\alpha) + \int_{N_0} d(\frac{\partial}{\partial t} \lrcorner \Lambda)$$
$$= \int_{N_0} g^\alpha \Psi_\alpha,$$

where in the last step we used the fixed-boundary condition, the vanishing of $F_0^* \theta^\alpha$, and the definition $g^\alpha = G^\alpha|_{N_0}$. Now the same reasoning as in §1.2 shows that $F_0 : N \hookrightarrow M$ is stationary for $\mathcal{F}_\Lambda$ under all fixed-boundary variations if and only if $\Psi_\alpha|_{N_0} = 0$ for all $\alpha = 1, \ldots, s$.

We now have a differential system $\{\theta^\alpha, d\theta^\alpha, \Psi_\alpha\}$ whose integral manifolds are exactly the integral manifolds of $\mathcal{I}$ that are stationary for $[\Lambda]$, but it is not clear that this system is uniquely determined by $[\Lambda]$ alone; we might get different systems for different admissible liftings. To rule out this possibility, observe first that if Λ, Λ' are any two admissible liftings of $[\Lambda]$, then the condition $\Lambda - \Lambda' \in \{I\}$ allows us to write $\Lambda - \Lambda' = \theta^\alpha \wedge \gamma_\alpha$, and then the fact that $d\Lambda \equiv d\Lambda' \equiv 0 \pmod{\{I\}}$ along with the structure equations (4.24) allows us to write

$$0 \equiv d\theta^\alpha \wedge \gamma_\alpha \equiv -\pi_i^\alpha \wedge \omega^i \wedge \gamma_\alpha \pmod{\{I\}}.$$

When $n \ge 2$, this implies that $\gamma_\alpha \equiv 0 \pmod{\{I\}}$, so while two general representatives of $[\Lambda]$ need be congruent only modulo $\{I\}$, for admissible liftings we have the following.

Proposition 4.5 *Two admissible liftings of the same $[\Lambda] \in \Gamma(\bigwedge^n (J/I))$ are congruent modulo $\{\bigwedge^2 I\}$.*

Of course, when $s = 1$, $\bigwedge^2 I = 0$ and we have a unique lifting, whose derivative is the familiar Poincaré-Cartan form. This explains how the Poincaré-Cartan form occurs in the context of point transformation as well as contact transformations.

We use the proposition as follows. If we take two admissible liftings Λ, Λ' of the same functional $[\Lambda]$, and write

$$\Lambda - \Lambda' = \tfrac{1}{2}\theta^\alpha \wedge \theta^\beta \wedge \gamma_{\alpha\beta},$$

with $\gamma_{\alpha\beta} + \gamma_{\beta\alpha} = 0$, then

$$\theta^\alpha \wedge (\Psi_\alpha - \Psi'_\alpha) = d(\Lambda - \Lambda') \equiv -\theta^\alpha \wedge d\theta^\beta \wedge \gamma_{\alpha\beta} \quad (\text{mod } \{\textstyle\bigwedge^2 I\}).$$

A consequence of this is that $\Psi_\alpha - \Psi'_\alpha \in \mathcal{I}$ for each α, and we can therefore give the following.

Definition 4.6 *The* Euler-Lagrange system $\mathcal{E}_\Lambda$ *of* $[\Lambda] \in \Gamma(M, \bigwedge^n(J/I))$ *is the differential ideal on M generated by I and the n-forms* $\{\Psi_1, \ldots, \Psi_s\} \subset \Omega^n(M)$, *where Λ is any admissible lifting of $[\Lambda]$ and $d\Lambda = \sum \theta^\alpha \wedge \Psi_\alpha$. A* stationary Legendre submanifold *of $[\Lambda]$ is an integral manifold of $\mathcal{E}_\Lambda$.*

4.2.3   The Betounes and Poincaré-Cartan Forms

For scalar variational problems, the Poincaré-Cartan form $\Pi \in \Omega^{n+1}(M)$ on the contact manifold (M, I) is an object of central importance. Some of its key features were outlined above. Underlying its usefulness is the fact that we are associating to a Lagrangian functional—a certain equivalence class of differential forms—an object that is not merely an equivalence class, but an actual differential form with which we can carry out certain explicit computations. We would like to construct an analogous object in the multi-contact case.

We will do this by imposing pointwise algebraic conditions on $\Pi \overset{def}{=} d\Lambda$. Fix an admissible coframing $(\theta^\alpha, \omega^i, \pi_i^\alpha)$ on a multi-contact manifold as in Definition 4.3. Then any admissible lifting $\Lambda \in \Gamma(\bigwedge^n J)$ of a functional $[\Lambda]$ has the form

$$\Lambda = \sum_{k=0}^{min(n,s)} \left((k!)^{-2} \sum_{|A|=|I|=k} F_A^I \theta^A \wedge \omega_{(I)} \right),$$

for some functions F_I^A, which are skew-symmetric with respect to each set of indices. Because J is integrable, $\Pi = d\Lambda$ lies in $\Omega^{n+1}(M) \cap \{\bigwedge^n J\}$; the "highest weight" part can be written as

$$\Pi = \sum_{k=1}^{min(n,s)} \left((k!)^{-2} \sum_{\substack{\alpha,i \\ |A|=|I|=k}} H_{\alpha A}^{iI} \pi_i^\alpha \wedge \theta^A \wedge \omega_{(I)} \right) \quad (\text{mod } \{\textstyle\bigwedge^{n+1} J\}).$$

$$(4.28)$$

The functions $H_{\alpha A}^{iI}$ are skew-symmetric in the multi-indices I and A. Notice that the equation $d\Pi \equiv 0 \pmod{\{\bigwedge^{n+1} J\}}$ gives for the $k = 1$ term

$$H_{\alpha\beta}^{ij} = H_{\beta\alpha}^{ji}.$$

To understand the relevant linear algebra, suppose that V^n is a vector space with basis $\{v_i\}$, and that W^s is a vector space with basis $\{w_\alpha\}$ and dual basis $\{w^\alpha\}$. Then we have for $k \geq 2$ the $GL(W) \times GL(V)$-equivariant exact sequence

$$0 \to U_k \to W^* \otimes V \otimes (\textstyle\bigwedge^k W^* \otimes \bigwedge^k V) \xrightarrow{\tau_k} \bigwedge^{k+1} W^* \otimes \bigwedge^{k+1} V \to 0.$$

Here the surjection is the obvious skew-symmetrization map, and U_k is by definition its kernel. The term $k = 1$ will be exceptional, and we instead define

$$0 \to U_1 \to \mathrm{Sym}^2(W^* \otimes V) \xrightarrow{\tau} \textstyle\bigwedge^2 W^* \otimes \bigwedge^2 V \to 0,$$

so that $U_1 = \mathrm{Sym}^2 W^* \otimes \mathrm{Sym}^2 V$.

Now we can regard our coefficients $H_{\alpha A}^{iI}$, with $|I| = |A| = k$, at each point of M as coefficients of an element

$$H_k = H_{\alpha A}^{iI} w^\alpha \otimes v_i \otimes w^A \otimes v_I \in W^* \otimes V \otimes (\textstyle\bigwedge^k W^* \otimes \bigwedge^k V).$$

Definition 4.7 *The form* $\Pi \in \Omega^{n+1}(M) \cap \{\bigwedge^n J\}$ *is* symmetric *if its expansion* (4.28) *has* $H_k \in U_k$ *for all* $k \geq 1$.

For $k = 1$ the condition is

$$H_{\alpha\beta}^{ij} = H_{\alpha\beta}^{ji} = H_{\beta\alpha}^{ij}.$$

We first need to show that the condition that a given Π be symmetric is independent of the choice of admissible coframe. Equivalently, we can show that the symmetry condition is preserved under the group of coframe changes of the form (4.23), and we will show this under three subgroups generating the group. First, it is obvious that a change

$$\bar{\theta}^\alpha = a_\beta^\alpha \theta^\beta, \quad \bar{\omega}^i = b_j^i \omega^j, \quad \bar{\pi}_i^\alpha = a_\beta^\alpha \pi_j^\beta (b^{-1})_i^j,$$

preserves the symmetry condition, because of the equivariance of the preceding exact sequences under $(a_\beta^\alpha) \times (b_j^i) \in GL(W) \times GL(V)$. Second, symmetry is preserved under

$$\bar{\theta}^\alpha = \theta^\alpha, \quad \bar{\omega}^i = \omega^i, \quad \bar{\pi}_i^\alpha = d_{i\beta}^\alpha \theta^\beta + e_{ij}^\alpha \omega^j + \pi_i^\alpha,$$

with $e_{ij}^\alpha = e_{ji}^\alpha$, because such a change has no effect on the expression for Π modulo $\{\bigwedge^{n+1} J\}$. Finally, consider a change of the form

$$\bar{\theta}^\alpha = \theta^\alpha, \quad \bar{\omega}^i = c_\beta^i \theta^\beta + \omega^i, \quad \bar{\pi}_i^\alpha = \pi_i^\alpha.$$

We will prove the invariance of the symmetry condition infinitesimally, writing instead of $\bar{\omega}^i$ the family

$$\omega^i(\varepsilon) = \varepsilon c_\beta^i \theta^\beta + \omega^i. \tag{4.29}$$

This associates to each H_k a tensor $H_{k+1}(\varepsilon)$ for each $k \geq 1$, and we will show that $\frac{d}{d\varepsilon}|_{\varepsilon=0} H_{k+1}(\varepsilon) \in U_{k+1}$. This just amounts to looking at the terms linear in ε when (4.29) is substituted into (4.28). Noting that $C = (c_\beta^i) \in W^* \otimes V$, we consider the commutative diagram

$$
\begin{array}{ccc}
W^* \otimes V \otimes \bigwedge^k W^* \otimes \bigwedge^k V \otimes W^* \otimes V & \xrightarrow{\sigma} & W^* \otimes V \otimes \bigwedge^{k+1} W^* \otimes \bigwedge^{k+1} V \\
\downarrow \tau_k \otimes 1 & & \downarrow \tau_{k+1} \\
\bigwedge^{k+1} W^* \otimes \bigwedge^{k+1} V \otimes W^* \otimes V & \rightarrow & \bigwedge^{k+2} W^* \otimes \bigwedge^{k+2} V,
\end{array}
$$

where σ is skew-symmetrization with the latter $W^* \otimes V$, and $\tau_k \otimes 1$ is an extension of the earlier skew-symmetrization. The point is that given $H_k \otimes C$ in the upper-left space of this diagram,

$$\tfrac{d}{d\varepsilon}|_{\varepsilon=0} H_{k+1}(\varepsilon) = \sigma(H_k \otimes C).$$

So if we assume that $(H_k) \in U_k$, then $\tau_k(H_k) = 0$, so $\sigma(H_k \times C) \in U_{k+1}$, which is what we wanted to show.

This proves that the condition that the symmetry condition on $\Pi = d\Lambda$ is independent of the choice of adapted coframe. We can now state the following.

Theorem 4.1 *Given a functional* $[\Lambda] \in \Gamma(M, \bigwedge^n(J/I))$, *there is a unique admissible lifting* $\Lambda \in \Gamma(M, \bigwedge^n J)$ *such that* $\Pi = d\Lambda \in \Omega^{n+1}(M)$ *is symmetric.*

Proof. We inductively construct $\Lambda = \Lambda_0 + \Lambda_1 + \cdots + \Lambda_{min(n,s)}$, with each $\Lambda_i \in \{\bigwedge^i I\}$ chosen to eliminate the fully skew-symmetric part of

$$\Pi_{i-1} \stackrel{def}{=} d(\Lambda_0 + \cdots + \Lambda_{i-1}).$$

Initially, $\Lambda_0 = F\omega$ is the prescribed X-semibasic n-form modulo $\{I\}$. We know from the existence of admissible liftings that there is some $\Lambda_1 \in \{I\}$ such that $\Pi_1 \stackrel{def}{=} d(\Lambda_0 + \Lambda_1) \in \{I\}$; and we know from Proposition 4.5 that Λ_1 is uniquely determined modulo $\{\bigwedge^2 I\}$. Now let

$$\Pi_1 \equiv H_{\alpha\beta}^{ij} \pi_i^\alpha \wedge \theta^\beta \wedge \omega_{(j)} \quad (\mathrm{mod}\ \{\textstyle\bigwedge^2 I\} + \{\textstyle\bigwedge^{n+1} J\}).$$

If we add to $\Lambda_0 + \Lambda_1$ the I-quadratic term

$$\Lambda_2 = \tfrac{1}{2!^2} F_{\alpha\beta}^{ij} \theta^\alpha \wedge \theta^\beta \wedge \omega_{(ij)},$$

then the structure equation (4.24) shows that this alters the I-linear term Π_1 only by

$$H_{\alpha\beta}^{ij} \rightsquigarrow H_{\alpha\beta}^{ij} + F_{\alpha\beta}^{ij}.$$

Because $F_{\alpha\beta}^{ij} = -F_{\alpha\beta}^{ji} = -F_{\beta\alpha}^{ij}$, we see that $F_{\alpha\beta}^{ij}$ may be uniquely chosen so that the new $H_{\alpha\beta}^{ij}$ lies in U_1.

The inductive step is similar. Suppose we have $\Lambda_0 + \cdots + \Lambda_l \in \Gamma(\bigwedge^n J)$ such that $\Pi_l = d(\Lambda_0 + \cdots + \Lambda_l)$ is symmetric modulo $\{\bigwedge^{l-1} I\}$. Then the term of I-degree l is of the form

$$\Pi_l \equiv \tfrac{1}{l!^2} \sum H_{\alpha A}^{iI} \pi_i^\alpha \wedge \theta^A \wedge \omega_{(I)} \quad (\mathrm{mod}\ \{U_1\} + \cdots + \{U_{l-1}\} + \{\bigwedge^{l+1} I\}).$$

for some $H_{\alpha A}^{iI}$. There is a unique skew-symmetric term

$$\Lambda_{l+i} = \tfrac{1}{(l+1)!^2} \sum_{|I|=|A|=l+1} F_A^I \theta^A \wedge \omega_{(I)}$$

which may be added so that

$$\Pi_{l+1} \in \{U_1\} + \cdots + \{U_l\} + \{\bigwedge^{l+1} I\}.$$

We can continue in this manner, up to $l = \min(n,s)$. $\qquad\square$

Definition 4.8 *The unique Λ in the preceding theorem is called the* Betounes *form for the functional* $[\Lambda]$.[4] *Its derivative* $\Pi = d\Lambda$ *is the* Poincaré-Cartan *form for* $[\Lambda]$.

The unique determination of Π, along with the invariance of the symmetry condition under admissible coframe changes of M, implies that Π is globally defined and invariant under symmetries of the functional $[\Lambda]$ and the multi-contact structure (M, I, J).

It is instructive to see the first step of the preceding construction in coordinates. If our initial Lagrangian is

$$\Lambda_0 = L(x, z, p)dx,$$

then we have already seen in (4.26) that

$$\Lambda_0 + \Lambda_1 = L\,dx + \theta^\alpha \wedge \tfrac{\partial L}{\partial p_i^\alpha} dx_{(i)}.$$

The H_1-term of $d(\Lambda_0 + \Lambda_1)$ (see (4.27)) is

$$\frac{\partial^2 L}{\partial p_i^\alpha \partial p_j^\beta} dp_j^\beta \wedge \theta^\alpha \wedge dx_{(i)}. \tag{4.30}$$

Of course $L_{p_i^\alpha p_j^\beta} = L_{p_j^\beta p_i^\alpha}$, corresponding to the fact that $H_1 \in \mathrm{Sym}^2(W^* \otimes V)$ automatically. The proof shows that we can add $\Lambda_2 \in \{\bigwedge^2 I\}$ so that Π_2 instead includes

$$\tfrac{1}{2}(L_{p_i^\alpha p_j^\beta} + L_{p_j^\alpha p_i^\beta}) \pi_i^\alpha \wedge \theta^\beta \wedge \omega_{(j)} \in U_1 = \mathrm{Sym}^2(W^*) \otimes \mathrm{Sym}^2 V.$$

---

[4] *It was introduced in coordinates in [Bet84], and further discussed in [Bet87].*

In fact, this corresponds to the *principal symbol* of the Euler-Lagrange PDE system (4.20), given by the symmetric $s \times s$ matrix

$$
\begin{aligned}
H_{\alpha\beta}(\xi) &= \frac{\partial^2 L}{\partial p_i^\alpha \partial p_j^\beta} \xi_i \xi_j \\
&= \tfrac{1}{2}(L_{p_i^\alpha p_j^\beta} + L_{p_j^\alpha p_i^\beta})\xi_i \xi_j, \qquad \xi \in V^*.
\end{aligned}
$$

In light of this, it is not surprising to find that only the symmetric part of (4.30) has invariant meaning.

Note also that if Lagrangians Λ, Λ' differ by a divergence,

$$
\Lambda - \Lambda' = d\lambda, \qquad \lambda \in \Gamma(M, \textstyle\bigwedge^{n-1} J),
$$

then the construction in the proof of Theorem 4.1 shows that the Poincaré-Cartan forms are equal, though the Betounes forms may not be. A related but more subtle property is the following.

Theorem 4.2 *For a functional* $[\Lambda] \in \Gamma(M, \bigwedge^n(J/I))$, *the Poincaré-Cartan form* $\Pi = 0$ *if and only if the Euler-Lagrange system is trivial,* $\mathcal{E}_\Lambda = \mathcal{I}$.

Proof. One direction is clear: if $\Pi = 0$, then the n-form generators Ψ_α for $\mathcal{E}_\Lambda$ can be taken to be 0, so that $\mathcal{E}_\Lambda = \mathcal{I}$. For the converse, we first consider the I-linear term

$$
\Pi_1 \equiv H_{\alpha\beta}^{ij}\pi_i^\alpha \wedge \theta^\beta \wedge \omega_{(j)} \quad (\mathrm{mod}\ \{\textstyle\bigwedge^2 I\} + \{\textstyle\bigwedge^{n+1} J\}).
$$

$\mathcal{E}_\Lambda$ is generated by $\mathcal{I}$ and $\Psi_\beta = H_{\alpha\beta}^{ij}\pi_i^\alpha \wedge \omega_{(j)}$, $1 \le \beta \le s$, and our assumption $\mathcal{E}_\Lambda = \mathcal{I}$ then implies that these $\Psi_\beta = 0$; that is,

$$
H_1 = H_{\alpha\beta}^{ij}\pi_i^\alpha \wedge \theta^\beta \wedge \omega_{(j)} = 0.
$$

We will first show that this implies

$$
H_2 = H_3 = \cdots = 0
$$

as well, which will imply $\Pi \in \{\bigwedge^{n+1} J\}$. To see this, suppose H_l is the first non-zero term, having I-degree l. Then we can consider

$$
0 \equiv d\Pi \quad (\mathrm{mod}\ \{\textstyle\bigwedge^l I\} + \{\textstyle\bigwedge^{n+1} J\}),
$$

and using the structure equations (4.24),

$$
0 = \sum_{|I|=|A|=l-1} H_{\alpha\beta A}^{ijI}\pi_i^\alpha \wedge \pi_j^\beta.
$$

Written out fully, this says that

$$
H_{\alpha_1\alpha_2\cdots\alpha_{l+1}}^{i_1 i_2\cdots i_{l+1}} = H_{\alpha_2\alpha_1\cdots\alpha_{l+1}}^{i_2 i_1\cdots i_{l+1}}.
$$

Also, $H^{iI}_{\alpha A}$ is fully skew-symmetric in I and A. But together, these imply that H_l is fully skew-symmetric in all upper and all lower indices, for

$$
\begin{aligned}
H^{i_1 i_2 i_3 \cdots}_{\alpha_1 \alpha_2 \alpha_3 \cdots} &= H^{i_2 i_1 i_3 \cdots}_{\alpha_2 \alpha_1 \alpha_3 \cdots} \\
&= -H^{i_2 i_1 i_3 \cdots}_{\alpha_2 \alpha_3 \alpha_1 \cdots} \\
&= -H^{i_1 i_2 i_3 \cdots}_{\alpha_3 \alpha_2 \alpha_1 \cdots} \\
&= H^{i_1 i_2 i_3 \cdots}_{\alpha_3 \alpha_1 \alpha_2 \cdots} \\
&= H^{i_2 i_1 i_3 \cdots}_{\alpha_1 \alpha_3 \alpha_2 \cdots} \\
&= -H^{i_2 i_1 i_3 \cdots}_{\alpha_1 \alpha_2 \alpha_3 \cdots}.
\end{aligned}
$$

This proves full skew-symmetry in the upper indices, and the proof for lower indices is similar. However, we constructed Π so that each H_k lies in the invariant complement of the fully skew-symmetric tensors, so we must have $H_k = 0$.

Now we have shown that if the Euler-Lagrange equations of $[\Lambda]$ are trivial, then $\Pi \in \{\bigwedge^{n+1} J\}$. But that means that the Betounes form Λ is not merely semibasic over the quotient space X, but actually basic. We can then compute the (assumed trivial) first variation down in X instead of M, and find that for any submanifold $N \hookrightarrow X$, and any vector field v along N vanishing at ∂N,

$$
0 = \int_N v \lrcorner \, d\Lambda.
$$

But this implies that $d\Lambda = 0$, which is what we wanted to prove.   □

The preceding results indicate that Π is a good generalization of the classical Poincaré-Cartan form for second-order, scalar Euler-Lagrange equations. We note that for higher-order Lagrangian functionals on vector-valued functions of one variable (i.e., functionals on curves), such a generalization is known, and not difficult; but for functionals of order $k \geq 2$ on vector-valued functions of several variables, little is known.[5]

We want to briefly mention a possible generalization to the multi-contact case of Noether's theorem, which gives an isomorphism from a Lie algebra of symmetries to a space of conservation laws. To avoid distracting global considerations, we will assume that $H^q_{dR}(M) = 0$ in all degrees $q > 0$. First, we have the space $\mathfrak{g}_\Pi$, consisting of vector fields on M which preserve I and Π,

$$
\mathfrak{g}_\Pi = \{v \in \mathcal{V}(M) : \mathcal{L}_v I \subseteq I, \ \mathcal{L}_v \Pi = 0\}.
$$

Second, we have the space of conservation laws

$$
\mathcal{C} = H^{n-1}(\Omega^*(M)/\mathcal{E}_\Lambda);
$$

under our topological assumption, this is identified with $H^n(\mathcal{E}_\Lambda)$, and we need not introduce a notion of "proper" conservation law as in §1.3. In this situation, Noether's theorem says the following.

---

[5] But see [Gra00].

There is a map $\eta : \mathfrak{g}_\Pi \to H^n(\mathcal{E}_\Lambda)$, defined by $v \mapsto [v \lrcorner \Pi]$, which is an isomorphism if Π is non-degenerate in a suitable sense.

The map is certainly well-defined; that is, for any $v \in \mathfrak{g}_\Pi$, the form $v \lrcorner \Pi$ is a closed section of $\mathcal{E}_\Lambda$. First,

$$v \lrcorner \Pi = (v \lrcorner \theta^\alpha)\Psi_\alpha - \theta^\alpha \wedge (v \lrcorner \Psi_\alpha),$$

so that $v \lrcorner \Pi$ is a section of $\mathcal{E}_\Lambda$; and second,

$$d(v \lrcorner \Pi) = \mathcal{L}_v \Pi - v \lrcorner d\Pi = 0,$$

so that $v \lrcorner \Pi$ is closed. However, the proof that under the right conditions this map is an isomorphism involves some rather sophisticated commutative algebra, generalizing the symplectic linear algebra used in Chapter 1. This will not be presented here.

As in the scalar case, a simple prescription for the conserved density in $H^{n-1}(\Omega^*(M)/\mathcal{E}_\Lambda)$ corresponding to $v \in \mathfrak{g}_\Pi$ is available when also

$$\mathcal{L}_v \Lambda = 0.$$

One virtue of the Betounes form is that this holds for infinitesimal multi-contact symmetries of $[\Lambda]$. Assuming only that $d\Lambda = \Pi$ and $\mathcal{L}_v\Lambda = 0$, we can calculate that

$$d(-v \lrcorner \Lambda) = -\mathcal{L}_v\Lambda + v \lrcorner d\Lambda = v \lrcorner \Pi. \tag{4.31}$$

Therefore, $-v \lrcorner \Lambda \in \Omega^{n-1}(M)$ represents a class in $\mathcal{C} = H^{n-1}(\Omega^*(M)/\mathcal{E}_\Lambda)$ corresponding to $\eta(v) \in H^n(\mathcal{E}_\Lambda)$. We will use this prescription in the following.

4.2.4 Harmonic Maps of Riemannian Manifolds

The most familiar variational PDE systems in differential geometry are those describing harmonic maps between Riemannian manifolds.

Let P, Q be Riemannian manifolds of dimensions n, s. We will define a Lagrangian density on P, depending on a map $P \to Q$ and its first derivatives, whose integral over P may be thought of as the *energy* of the map. The appropriate multi-contact manifold for this is the space of 1-jets of maps $P \to Q$,

$$M = J^1(P, Q),$$

whose multi-contact system will be described shortly. We may also think of M as $\mathrm{Hom}(TP, TQ)$, the total space of a rank-ns vector bundle over $P \times Q$. To carry out computations, it will be most convenient to work on

$$\mathcal{F} \overset{\text{def}}{=} \mathcal{F}(P) \times \mathcal{F}(Q) \times \mathbf{R}^{ns},$$

where $\mathcal{F}(P)$, $\mathcal{F}(Q)$ are the orthonormal frame bundles. These are parallelized in the usual manner by (ω^i, ω^i_j), $(\varphi^\alpha, \varphi^\alpha_\beta)$, respectively, with structure equations

$$\begin{cases} d\omega^i = -\omega^i_j \wedge \omega^j, & d\omega^i_j = -\omega^i_k \wedge \omega^k_j + \Omega^i_j, \\ d\varphi^\alpha = -\varphi^\alpha_\beta \wedge \varphi^\beta, & d\varphi^\alpha_\beta = -\varphi^\alpha_\gamma \wedge \varphi^\gamma_\beta + \Phi^\alpha_\beta. \end{cases}$$

These forms and structure equations will be considered pulled back to $\mathcal{F}$. To complete a coframing of $\mathcal{F}$, we take linear fiber coordinates p_i^α on $\mathbf{R}^{ns}$, and define

$$\pi_i^\alpha = dp_i^\alpha + \varphi_\beta^\alpha p_i^\beta - p_j^\alpha \omega_i^j.$$

The motivation here is that $\mathrm{Hom}(TP, TQ) \to P \times Q$ is a vector bundle associated to the principal $(O(n) \times O(s))$-bundle $\mathcal{F}(P) \times \mathcal{F}(Q) \to P \times Q$, with the data $((e_i^P), (e_\alpha^Q), (p_i^\alpha)) \in \mathcal{F}$ defining the homomorphism $e_i^P \mapsto e_\alpha^Q p_i^\alpha$. Furthermore, if a section $\sigma \in \Gamma(\mathrm{Hom}(TP, TQ))$ is represented by an equivariant map (p_i^α) : $\mathcal{F}(P) \times \mathcal{F}(Q) \to \mathbf{R}^{ns}$, then the $\mathbf{R}^{ns}$-valued 1-form (π_i^α) represents the covariant derivative of σ.

For our purposes, note that $M = \mathrm{Hom}(TP, TQ)$ is the quotient of $\mathcal{F}$ under a certain action of $O(n) \times O(s)$, and that the forms semibasic for the projection $\mathcal{F} \to M$ are generated by $\omega^i, \varphi^\alpha, \pi_i^\alpha$. A natural multi-contact system on M pulls back to $\mathcal{F}$ as the Pfaffian system I generated by

$$\theta^\alpha \overset{def}{=} \varphi^\alpha - p_i^\alpha \omega^i,$$

and the associated integrable Pfaffian system on M pulls back to $J = \{\varphi^\alpha, \omega^i\} = \{\theta^\alpha, \omega^i\}$. The structure equations on $\mathcal{F}$ adapted to these Pfaffian systems are

$$\begin{cases} d\theta^\alpha = -\pi_i^\alpha \wedge \omega^i - \varphi_\beta^\alpha \wedge \theta^\beta, \\ d\omega^i = -\omega_j^i \wedge \omega^j, \\ d\pi_i^\alpha = \Phi_\beta^\alpha p_i^\beta - p_j^\alpha \Omega_i^j - \varphi_\beta^\alpha \wedge \pi_i^\beta - \pi_j^\alpha \wedge \omega_i^j. \end{cases} \quad (4.32)$$

We now define the *energy Lagrangian*

$$\tilde{\Lambda} = \tfrac{1}{2}||p||^2 \omega \in \Gamma(\textstyle\bigwedge^n J) \subset \Omega^n(\mathcal{F}),$$

where the norm is

$$||p||^2 = \mathrm{Tr}(p^* p) = \sum (p_i^\alpha)^2.$$

Although this $\tilde{\Lambda}$ is not an admissible lifting of its induced functional $[\Lambda]$, a computation using the structure equations (4.32) shows that

$$\Lambda \overset{def}{=} \tfrac{1}{2}||p||^2 \omega + p_i^\alpha \theta^\alpha \wedge \omega_{(i)}$$

is admissible:

$$\begin{aligned} d\Lambda &= -\theta^\alpha \wedge \pi_i^\alpha \wedge \omega_{(i)} - p_i^\alpha p_i^\beta \varphi_\beta^\alpha \wedge \omega + p_i^\alpha p_j^\alpha \omega_i^j \wedge \omega \\ &= -\theta^\alpha \wedge \pi_i^\alpha \wedge \omega_{(i)}, \end{aligned}$$

where the last step uses $\varphi_\beta^\alpha + \varphi_\alpha^\beta = \omega_j^i + \omega_i^j = 0$. Now we define

$$\Pi = -\theta^\alpha \wedge \pi_i^\alpha \wedge \omega_{(i)},$$

and note that Π is in fact the lift to $\mathcal{F}$ of a *symmetric* form on M, as defined earlier. Therefore, we have found the Betounes form and the Poincaré-Cartan form for the energy functional.

The Euler-Lagrange system for $[\Lambda]$, pulled back to $\mathcal{F}$, is

$$\mathcal{E}_\Lambda = \{\theta^\alpha,\ \pi_i^\alpha \wedge \omega^i,\ \pi_i^\alpha \wedge \omega_{(i)}\}.$$

A Legendre submanifold $N \hookrightarrow M = J^1(P,Q)$ on which $\bigwedge \omega^i \neq 0$ is the 1-jet graph of a map $f : P \to Q$. On the inverse image $\pi^{-1}(N) \subset \mathcal{F}$, in addition to $\theta^\alpha = 0$, there are relations

$$\pi_i^\alpha = h_{ij}^\alpha \omega^j, \quad h_{ij}^\alpha = h_{ji}^\alpha.$$

Differentiating this equation shows that the expression

$$h = h_{ij}^\alpha \omega^i \omega^j \otimes e_\alpha^Q$$

is invariant along fibers of $\pi^{-1}(N) \to N$, so it gives a well-defined section of $\mathrm{Sym}^2(T^*P) \otimes TQ$; this is called the *second fundamental form* of the map $f : P \to Q$. The condition for N to be an integral manifold of the Euler-Lagrange system is then

$$\mathrm{Tr}(h) = h_{ii}^\alpha = 0 \in \Gamma(M, f^*TQ).$$

Definition 4.9 *A map $f : P \to Q$ between Riemannian manifolds is* harmonic *if the trace of its second fundemental form vanishes.*

Expressed in coordinates on P and Q, this is a second-order PDE system for $f : P \to Q$.

We now consider conservation laws for the harmonic map system $\mathcal{E}_\Lambda \subset \Omega^*(\mathcal{F})$ corresponding to infinitesimal isometries (Killing vector fields of either P or Q). These are symmetries not only of Π but of the Lagrangian Λ, so we can use the simplified prescription (4.31) for a conserved $(n-1)$-form.

First, an infinitesimal isometry of P induces a unique vector field on $M = J^1(P,Q)$ preserving I and fixing Q. This vector field preserves Λ, Π, and $\mathcal{E}_\Lambda$, and has a natural lift to $\mathcal{F}$ which does the same. This vector field v on $\mathcal{F}$ satisfies

$$v \lrcorner\, \omega^i = v^i, \quad v \lrcorner\, \varphi^\alpha = 0 \quad \Rightarrow \quad v \lrcorner\, \theta^\alpha = -p_i^\alpha v^i,$$

for some functions v^i. We can then calculate

$$\varphi_v \overset{def}{=} v \lrcorner\, \Lambda \equiv (\tfrac{1}{2}||p||^2 v^i - p_i^\alpha p_j^\alpha v^j)\omega_{(i)} \quad (\mathrm{mod}\ \{I\}).$$

As in Chapter 3, it is useful to write this expression restricted to the 1-jet graph of a map $f : P \to Q$, which is

$$
\begin{aligned}
v \lrcorner\, \Lambda|_N &= *_P(\tfrac{1}{2}||p||^2 v^i - p_i^\alpha p_j^\alpha v^j)\omega^i \\
&= *_P \tfrac{1}{2}\left(v \lrcorner\, (\tfrac{1}{2}||df||^2 \textstyle\sum(\omega^i)^2 - f^* \textstyle\sum(\varphi^\alpha)^2)\right),
\end{aligned}
$$

where we use $f^*\varphi^\alpha = p_i^\alpha \omega^i$, and $*_P\omega_{(i)} = \omega^i$. One might recognize the *stress-energy tensor*

$$S = \tfrac{1}{2}||df||^2 ds_P^2 - f^* ds_Q^2,$$

and write our conserved density as

$$2(v \lrcorner \Lambda) \equiv *_P(v \lrcorner S) \quad (\mathrm{mod}\ \{I\}). \tag{4.33}$$

In fact, S is traditionally defined as the unique symmetric 2-tensor on P for which the preceding equation holds for arbitrary $v \in \mathcal{V}(P)$ and $f : P \to Q$; then (4.33) gives a conserved density when v is an infinitesimal isometry and f is a harmonic map. In fact, for any infinitesimal isometry v, a calculation gives

$$d(*_P(v \lrcorner S)) = (v \lrcorner \mathrm{div}\ S)\omega \tag{4.34}$$

on the 1-jet graph of any map.[6]

Now consider an infinitesimal isometry of Q, whose lift $w \in \mathcal{V}(\mathcal{F})$ satisfies

$$w \lrcorner \omega^i = 0, \ w \lrcorner \varphi^\alpha = w^i \quad \Rightarrow \quad w \lrcorner \theta^\alpha = w^\alpha.$$

Then

$$\varphi_w \overset{def}{=} w \lrcorner \Lambda = w^\alpha p_i^\alpha \omega_{(i)}.$$

Given a map $f : P \to Q$, we can use $df \in \mathrm{Hom}(TP, TQ)$ and $ds_Q^2 \in \mathrm{Sym}^2(T^*Q)$ to regard $ds_Q^2(df(\cdot), w)$ as a 1-form on P, and then

$$\varphi_w = *_P(ds_Q^2(df(\cdot), w)).$$

Because this expression depends linearly on w, we can simplify further by letting $\mathfrak{a}$ denote the Lie algebra of infinitesimal symmetries of Q, and then the map $w \mapsto w \lrcorner \Lambda$ is an element of $\mathfrak{a}^* \otimes \Omega^{n-1}(P)$. If we define an $\mathfrak{a}^*$-valued 1-form on P by

$$\alpha(v) = ds_Q^2(df(v), \cdot), \quad v \in T_pP,$$

then our conservation laws read

$$d(*_P\alpha) = 0 \in \mathfrak{a}^* \otimes \Omega^n(P). \tag{4.35}$$

The $\mathfrak{a}^*$-valued $(n - 1)$-form $*_P\alpha$ may be formed for any map $f : P \to Q$, and it is closed if f is harmonic. In fact, if Q is locally homogeneous, meaning that infinitesimal isometries span each tangent space T_qQ, then (4.35) is *equivalent* to the harmonicity of f.

An important special case of this last phenomenon is when Q itself is a Lie group G with bi-invariant metric ds_G^2. Examples are compact semisimple Lie groups, such as $O(N)$ or $SU(N)$, with metric induced by the Killing form on the Lie algebra $\mathfrak{g}$. Now a map $f : P \to G$ is uniquely determined up to left-translation by the pullback $f^*\varphi$ of the left-invariant $\mathfrak{g}$-valued Maurer-Cartan 1-form φ. Using the metric to identify $\mathfrak{g} \cong \mathfrak{g}^*$, the conservation laws state that

---

[6]The *divergence* of a symmetric 2-form S is the 1-form $\mathrm{div}\ S = \nabla_{e_i}S(e_i, \cdot)$, where ∇ is the Levi-Civita covariant derivative and (e_i) is any orthonormal frame. Equation (4.34) is true of any symmetric 2-form S and infinitesimal isometry v.

if f is harmonic, then $d(*_P(f^*\varphi)) = 0$. Conversely, if P is simply connected, then given a $\mathfrak{g}$-valued 1-form α on P satisfying

$$\begin{cases} d\alpha + \frac{1}{2}[\alpha, \alpha] = 0, \\ d(*_P\alpha) = 0, \end{cases}$$

there is a harmonic map $f : P \to G$ with $f^*\varphi = \alpha$, uniquely determined up to left-translation. This is the idea behind the gauge-theoretic reformulation of certain harmonic map systems, for which remarkable results have been obtained in the past decade.[7] Quite generally, PDE systems that can be written as systems of conservation laws have special properties; one typically exploits such expressions to define weak solutions, derive integral identities, and prove regularity theorems.

4.3   Higher-Order Conservation Laws

One sometimes encounters a conservation law for a PDE that involves higher-order derivatives of the unknown function, but that cannot be expressed in terms of derivatives of first-order conservation laws considered up to this point. An example is the $(1+1)$-dimensional wave equation $-z_{tt} + z_{xx} = 0$, for which $(z_{tt}^2 + z_{tx}^2)dt + 2z_{tt}z_{tx}dx$ is closed on solutions, but cannot be obtained by differentiating any conservation law on $J^1(\mathbf{R}^2, \mathbf{R})$. In this section, we introduce the geometric framework in which such conservation laws may be found, and we propose a version of Noether's theorem appropriate to this setting. While other general forms of Noether's theorem have been stated and proved (e.g., see [Vin84] or [Olv93]), it is not clear how they relate to that conjectured here.

We also discuss (independently from the preceding) the higher-order relationship between surfaces in Euclidean space with Gauss curvature $K = -1$ and the sine-Gordon equation $z_{tx} = \frac{1}{2}\sin(2z)$, in terms of exterior differential systems.

4.3.1   The Infinite Prolongation

We begin by defining the *prolongation* of an exterior differential system (EDS). When this is applied to the EDS associated to a PDE system, it gives the EDS associated to the PDE system augmented by the first derivatives of the original equations. This construction then extends to that of the *infinite prolongation*, an EDS on an infinite-dimensional manifold which includes information about derivatives of all orders.

The general definition of prolongation uses a construction introduced in §4.2, in the discussion of multi-contact manifolds. Let X^{n+s} be a manifold, and $G_n(TX) \xrightarrow{\pi} X$ the bundle of tangent n-planes of X; points of $G_n(TX)$ are of the form (p, E), where $p \in X$ and $E \subset T_pX$ is a vector subspace of dimension

---

[7]The literature on this subject is vast, but a good starting point is [Woo94].

n. As discussed previously, there is a canonical Pfaffian system $I \subset T^*G_n(TX)$ of rank s, defined at (p, E) by

$$I_{(p,E)} \overset{def}{=} \pi^*(E^\perp).$$

Given local coordinates (x^i, z^α) on X, there are induced coordinates $(x^i, z^\alpha, p_i^\alpha)$ on $G_n(TX)$, in terms of which I is generated by the 1-forms

$$\theta^\alpha = dz^\alpha - p_i^\alpha dx^i. \tag{4.36}$$

We let $\mathcal{I} \subset \Omega^*(G_n(TX))$ be the differential ideal generated by I.

Now let $(M, \mathcal{E})$ be an exterior differential system; that is, M is a manifold of dimension $m+s$ and $\mathcal{E} \subset \Omega^*(M)$ is a differential ideal for which we are interested in m-dimensional integral manifolds. We then define the locus $M^{(1)} \subset G_m(TM)$ to consist of the *integral elements* of $\mathcal{E} \subset \Omega^*(M)$; that is, $(p, E) \in M^{(1)}$ if and only if

$$\varphi_E = 0 \in {\textstyle\bigwedge}^*(E^*) \quad \text{for all } \varphi \in \mathcal{E}.$$

We will assume from now on that $M^{(1)} \overset{\iota}{\hookrightarrow} G_m(TM)$ is a smooth submanifold. Then we define

$$\mathcal{E}^{(1)} \overset{def}{=} \iota^* \mathcal{I} \subset \Omega^*(M^{(1)})$$

as the restriction to $M^{(1)}$ of the multi-contact differential ideal. This is the same as the differential ideal generated by the Pfaffian system $\iota^* I \subset T^*M^{(1)}$, and the *first prolongation* of $(M, \mathcal{E})$ is defined to be the exterior differential system $(M^{(1)}, \mathcal{E}^{(1)})$. Note that the first prolongation is always a Pfaffian system. Furthermore, if $\pi : M^{(1)} \to M$ is the obvious projection map, and assuming that $\mathcal{E}$ is a Pfaffian system, then one can show that $\pi^*\mathcal{E} \subseteq \mathcal{E}^{(1)}$. However, the projection π could be quite complicated, and need not even be surjective. Finally, note that any integral manifold $f : N \hookrightarrow M$ of $\mathcal{E}$ lifts to an integral manifold $f^{(1)} : N \hookrightarrow M^{(1)}$ of $\mathcal{E}^{(1)}$, and that the transverse integral manifold of $\mathcal{E}^{(1)}$ is locally of this form.

Inductively, the *kth prolongation* $(M^{(k)}, \mathcal{E}^{(k)})$ of $(M, \mathcal{E})$ is the first prolongation of the $(k-1)$st prolongation of $(M, \mathcal{E})$. This gives rise to the *prolongation tower*

$$\cdots \to M^{(k)} \to M^{(k-1)} \to \cdots \to M^{(1)} \to M.$$

An integral manifold of $(M, \mathcal{E})$ lifts to an integral manifold of each $(M^{(k)}, \mathcal{E}^{(k)})$ in this tower.

Two examples will help to clarify the construction. The first is the prolongation tower of the multi-contact system $(G_n(TX), \mathcal{I})$ itself, and this will give us more detailed information about the structure of the ideals $\mathcal{E}^{(k)}$ for general $(M, \mathcal{E})$. The second is the prolongation tower of the EDS associated to a first-order PDE system, most of which we leave as an exercise.

Example 1. Consider the multi-contact ideal $\mathcal{I}$ on $G_n(TX)$, over a manifold X of dimension $n + s$ with local coordinates (x^i, z^α). We can see from the coordinate expression (4.36) that its integral elements over the dense open subset where $\bigwedge_i dx^i \neq 0$ are exactly the n-planes of the form

$$E_{p_{ij}^\alpha} = \{dz^\alpha - p_i^\alpha dx^i, \; dp_i^\alpha - p_{ij}^\alpha dx^j\}^\perp \subset T(G_n(TX)),$$

for some constants $p_{ij}^\alpha = p_{ji}^\alpha$. These p_{ij}^α are local fiber coordinates for the prolongation $(G_n(TX)^{(1)}, \mathcal{I}^{(1)})$. Furthermore, with respect to the full coordinates $(x^i, z^\alpha, p_i^\alpha, p_{ij}^\alpha)$ for $G_n(TX)^{(1)} \subset G_n(TG_n(TX))$, the 1-jet graphs of integral manifolds of $\mathcal{I} \subset \Omega^*(G_n(TX))$ satisfy

$$dz^\alpha - p_i^\alpha dx^i = 0, \; dp_i^\alpha - p_{ij}^\alpha dx^j = 0.$$

It is these $s + ns$ 1-forms that differentially generate the prolonged Pfaffian system $\mathcal{I}^{(1)}$. It is not difficult to verify that we have globally $G_n(TX)^{(1)} = G_{2,n}(X)$, the bundle of 2-jets of n-dimensional submanifolds of X, and that $\mathcal{I}^{(1)} \subset \Omega^*(G_{2,n}(X))$ is the Pfaffian system whose transverse integral manifolds are 2-jet graphs of submanifolds of X.

More generally, let $G_k = G_{k,n}(X) \to X$ be the bundle of k-jets of n-dimensional submanifolds of X. Because a 1-jet of a submanifold is the same as a tangent plane, $G_1 = G_n(TX)$ is the original space whose prolongation tower we are describing. G_k carries a canonical Pfaffian system $\mathcal{I}_k \subset \Omega^*(G_k)$, whose transverse integral manifolds are k-jet graphs $f^{(k)} : N \hookrightarrow G_k$ of n-dimensional submanifolds $f : N \hookrightarrow X$. This is perhaps clearest in coordinates. Letting (x^i, z^α) be coordinates on X, G_k has induced local coordinates $(x^i, z^\alpha, p_i^\alpha, \ldots, p_I^\alpha)$, $|I| \leq k$, corresponding to the jet at (x^i, z^α) of the submanifold

$$\{(\bar{x}^i, \bar{z}^\alpha) \in X : \bar{z}^\alpha = z^\alpha + p_i^\alpha(\bar{x}^i - x^i) + \cdots + \tfrac{1}{I!}p_I^\alpha(\bar{x} - x)^I\}.$$

In terms of these coordinates, the degree-1 part $I_k \subset T^*(G_k)$ of the Pfaffian system $\mathcal{I}_k$ is generated by

$$\begin{aligned}
\theta^\alpha &= dz^\alpha - p_i^\alpha dx^i, \\
\theta_i^\alpha &= dp_i^\alpha - p_{ij}^\alpha dx^j, \\
&\;\vdots \\
\theta_I^\alpha &= dp_I^\alpha - p_{Ij}^\alpha dx^j, \quad |I| = k - 1.
\end{aligned} \tag{4.37}$$

It is not hard to see that the transverse integral manifolds of this I_k are as described above. The point here is that $(G_k, \mathcal{I}_k)$ is the first prolongation of $(G_{k-1}, \mathcal{I}_{k-1})$ for each $k > 1$, and is therefore the $(k - 1)$st prolongation of the original $(G_1, \mathcal{I}_1) = (G_n(TX), \mathcal{I})$.

For future reference, we note the structure equations

$$\begin{aligned}
d\theta^\alpha &= -\theta_i^\alpha \wedge dx^i, \\
d\theta_i^\alpha &= -\theta_{ij}^\alpha \wedge dx^j, \\
&\;\vdots \\
d\theta_I^\alpha &= -\theta_{Ij}^\alpha \wedge dx^j, \quad |I| = k - 2, \\
d\theta_I^\alpha &= -dp_{Ij}^\alpha \wedge dx^j, \quad |I| = k - 1.
\end{aligned} \tag{4.38}$$

There is a tower

$$\cdots \to G_k \to G_{k-1} \to \cdots \to G_1, \qquad (4.39)$$

and one can pull back to G_k any functions or differential forms on $G_{k'}$, with $k' < k$. Under these maps, our different uses of the coordinates p_I^α and forms θ_I^α are consistent, and we can also write

$$I_{k'} \subset I_k \subset T^* G_k, \qquad \text{for } k' < k.$$

None of the I_k is an integrable Pfaffian system. In fact, the filtration on G_k

$$I_k \supset I_{k-1} \supset \cdots \supset I_1 \supset 0$$

coincides with the *derived flag* of $I_k \subset T^* G_{k,n}$ (cf. Ch. II, §4 of [B+91]).
Example 2. Our second example of prolongation relates to a first-order PDE system $F^a(x^i, z^\alpha(x), z_{x^i}^\alpha(x)) = 0$ for some unknown functions $z^\alpha(x)$. The equations $F^a(x^i, z^\alpha, p_i^\alpha)$ define a locus M_F in the space $J^1(\mathbf{R}^n, \mathbf{R}^s)$ of 1-jets of maps $z : \mathbf{R}^n \to \mathbf{R}^s$, and we will assume that this locus is a smooth submanifold which submersively surjects onto $\mathbf{R}^n$. The restriction to $M_F \subset J^1(\mathbf{R}^n, \mathbf{R}^s)$ of the multi-contact Pfaffian system $I_1 = \{dz^\alpha - p_i^\alpha dx^i\}$ generates an EDS $(M_F, \mathcal{I}_F)$. Now, the set of integral elements for $(M_F, \mathcal{I}_F)$ is a subset of the set of integral elements for $\mathcal{I}_1$ in $J^1(\mathbf{R}^n, \mathbf{R}^s)$; it consists of those integral elements of $\mathcal{I}_1$ which are tangent to $M_F \subset J^1(\mathbf{R}^n, \mathbf{R}^s)$. Just as in the preceding example, the integral elements of $\mathcal{I}_1$ may be identified with elements of the space $J^2(\mathbf{R}^n, \mathbf{R}^s)$ of 2-jets of maps. The collection of 2-jets which correspond to integral elements of (M_F, I_F) are exactly the 2-jets satisfying the augmented PDE system

$$\begin{aligned}
0 &= F^a(x^i, z^\alpha(x), z_{x^i}^\alpha(x)), \\
0 &= \frac{\partial F^a}{\partial x^i} + \frac{\partial F^a}{\partial z^\alpha} z_{x^i}^\alpha + \frac{\partial F^a}{\partial p_j^\alpha} z_{x^i x^j}^\alpha.
\end{aligned}$$

Therefore, integral manifolds of the prolongation of the EDS associated to a PDE system correspond to solutions of this augmented system. For this reason, prolongation may generally be thought of as adjoining the derivatives of the original equations.

It is important to note that as the first prolongation of arbitrary $(M, \mathcal{E})$ is embedded in the canonical multi-contact system $(G_n(TM), \mathcal{I})$, so can all higher prolongations $(M^{(k)}, \mathcal{E}^{(k)})$ be embedded in the prolongations $(G_{k,n}(M), \mathcal{I}_k)$. Among other things, this implies that $\mathcal{E}^{(k)}$ is locally generated by forms like (4.37), satisfying structure equations (4.38), typically with additional linear-algebraic relations.

Of most interest to us is the *infinite prolongation* $(M^{(\infty)}, \mathcal{E}^{(\infty)})$ of an EDS $(M, \mathcal{E})$. As a space, $M^{(\infty)}$ is defined as the inverse limit of

$$\cdots \overset{\pi_{k+1}}{\to} M^{(k)} \overset{\pi_k}{\to} \cdots \overset{\pi_1}{\to} M^{(0)} = M;$$

that is,

$$M^{(\infty)} = \{(p_0, p_1, \ldots) \in M^{(0)} \times M^{(1)} \times \cdots : \pi_k(p_k) = p_{k-1} \text{ for each } k \geq 1\}.$$

An element of $M^{(\infty)}$ may be thought of as a Taylor series expansion for a possible integral manifold of $(M, \mathcal{E})$. $M^{(\infty)}$ is generally of infinite dimension, but its presentation as an inverse limit will prevent us from facing analytic difficulties. In particular, smooth functions and differential forms are by definition the corresponding objects on some finite $M^{(k)}$, pulled up to $M^{(\infty)}$ by the projections. It therefore makes sense to define

$$\mathcal{E}^{(\infty)} = \bigcup_{k>0} \mathcal{E}^{(k)},$$

which gives an EDS on $M^{(\infty)}$ whose transverse integral manifolds are the infinite-jet graphs of integral manifolds of $(M^{(0)}, \mathcal{E}^{(0)})$. $\mathcal{E}^{(\infty)}$ is a Pfaffian system, differentially generated by its degree-1 part $I^{(\infty)} = \bigcup I^{(k)}$, where each $I^{(k)}$ is the degree-1 part of $\mathcal{E}^{(k)}$. In fact, we can see from (4.38) that $\mathcal{E}^{(\infty)}$ is *algebraically* generated by $I^{(\infty)}$; that is, $I^{(\infty)}$ is a formally integrable Pfaffian system, although this is not true of any finite $I^{(k)}$. However, there is no analog of the Frobenius theorem for the infinite-dimensional $M^{(\infty)}$, so we must be cautious about how we use this fact.

Vector fields on $M^{(\infty)}$ are more subtle. By definition, $\mathcal{V}(M^{(\infty)})$ is the Lie algebra of derivations of the ring $\mathcal{R}(M^{(\infty)})$ of smooth functions on $M^{(\infty)}$. In case $M^{(\infty)} = J^{\infty}(\mathbf{R}^n, \mathbf{R}^s)$, a vector field is of the form

$$v = v^i \frac{\partial}{\partial x^i} + v_0^\alpha \frac{\partial}{\partial z^\alpha} + \cdots + v_I^\alpha \frac{\partial}{\partial p_I^\alpha} + \cdots.$$

Each coefficient v_I^α is a function on some $J^k(\mathbf{R}^n, \mathbf{R}^s)$, possibly with $k > |I|$. Although v may have infinitely many terms, only finitely many appear in its application to any particular $f \in \mathcal{R}(M^{(\infty)})$, so there are no issues of convergence.

4.3.2 Noether's Theorem

To give the desired generalization of Noether's theorem, we must first discuss a generalization of the infinitesimal symmetries used in the classical version. For convenience, we change notation and let $(M, \mathcal{E})$ denote the infinite prolongation of an exterior differential system $(M^{(0)}, \mathcal{E}^{(0)})$.

Definition 4.10 *A* generalized symmetry *of* $(M^{(0)}, \mathcal{E}^{(0)})$ *is a vector field* $v \in \mathcal{V}(M)$ *such that* $\mathcal{L}_v \mathcal{E} \subseteq \mathcal{E}$. *A* trivial generalized symmetry *is a vector field* $v \in \mathcal{V}(M)$ *such that* $v \lrcorner \mathcal{E} \subseteq \mathcal{E}$. *The space* $\mathfrak{g}$ *of* proper generalized symmetries *is the quotient of the space of generalized symmetries by the subspace of trivial generalized symmetries.*

Several remarks are in order.

- The Lie derivative in the definition of generalized symmetry is defined by the Cartan formula

$$\mathcal{L}_v \varphi = v \lrcorner d\varphi + d(v \lrcorner \varphi).$$

  The usual definition involves a flow along v, which may not exist in this setting.

- A trivial generalized symmetry is in fact a generalized symmetry; this is an immediate consequence of the fact that $\mathcal{E}$ is differentially closed.

- The space of generalized symmetries has the obvious structure of a Lie algebra.

- Using the fact that $\mathcal{E}$ is a formally integrable Pfaffian system, it is easy to show that the condition $v \lrcorner \mathcal{E} \subseteq \mathcal{E}$ for v to be a trivial generalized symmetry is equivalent to the condition $v \lrcorner I = 0$, where $I = \mathcal{E} \cap \Omega^1(M)$ is the degree-1 part of $\mathcal{E}$.

- The vector subspace of trivial generalized symmetries is an ideal in the Lie algebra of generalized symmetries, so $\mathfrak{g}$ is a Lie algebra as well. The following proof of this fact uses the preceding characterization $v \lrcorner I = 0$ for trivial generalized symmetries: if $\mathcal{L}_v \mathcal{E} \subset \mathcal{E}$, $w \lrcorner I = 0$, and $\theta \in \Gamma(I)$, then

$$
\begin{aligned}
[v, w] \lrcorner \theta &= -w \lrcorner (v \lrcorner d\theta) + v(w \lrcorner \theta) - w(v \lrcorner \theta) \\
&= -w \lrcorner (\mathcal{L}_v \theta - d(v \lrcorner \theta)) + 0 - w \lrcorner d(v \lrcorner \theta) \\
&= -w \lrcorner \mathcal{L}_v \theta \\
&= 0.
\end{aligned}
$$

The motivation for designating certain generalized symmetries as trivial comes from a formal calculation which shows that a trivial generalized symmetry is tangent to any integral manifold of the formally integrable Pfaffian system $\mathcal{E}$. Thus, the "flow" of a trivial generalized symmetry does not permute the integral manifolds of $\mathcal{E}$, but instead acts by diffeomorphisms of each "leaf."

The following example is relevant to what follows. Let $M^{(0)} = J^1(\mathbf{R}^n, \mathbf{R})$ be the standard contact manifold of 1-jets of functions, with global coordinates (x^i, z, p_i) and contact ideal $\mathcal{E}^{(0)} = \{dz - p_i dx^i, \, dp_i \wedge dx^i\}$. The infinite prolongation of $(M^{(0)}, \mathcal{E}^{(0)})$ is

$$
M = J^\infty(\mathbf{R}^n, \mathbf{R}), \quad \mathcal{E} = \{\theta_I : |I| \geq 0\},
$$

where M has coordinates $(x^i, z, p_i, p_{ij}, \ldots)$, and $\theta_I = dp_I - p_{Ij} dx^j$. (For the empty index $I = \emptyset$, we let $p = z$, so $\theta = dp - p_i dx^i$ is the original contact form.) Then the trivial generalized symmetries of $(M, \mathcal{E})$ are the total derivative vector fields

$$
D_i = \tfrac{\partial}{\partial x^i} + p_i \tfrac{\partial}{\partial z} + \cdots + p_{Ii} \tfrac{\partial}{\partial p_I} + \cdots.
$$

We will determine the proper generalized symmetries of $(M, \mathcal{E})$ shortly.

There is another important feature of a vector field on the infinite prolongation $(M, \mathcal{E})$ of $(M^{(0)}, \mathcal{E}^{(0)})$, which is its *order*. To introduce this, first note that any vector field $v_0 \in \mathcal{V}(M^{(0)})$ on the original, finite-dimensional manifold induces a vector field and a flow on each finite prolongation $M^{(k)}$, and therefore induces on M itself a vector field $v \in \mathcal{V}(M)$ having a flow. A further special property of $v \in \mathcal{V}(M)$ induced by $v_0 \in \mathcal{V}(M^{(0)})$ is that $\mathcal{L}_v(I_k) \subseteq I_k$ for each

$k \geq 1$. Though it is tempting to try to characterize those $v \in \mathcal{V}(M)$ induced by such v_0 using this last criterion, we ought not to do so, because this is not a criterion that can be inherited by *proper* generalized symmetries of $(M, \mathcal{E})$. Specifically, an arbitrary trivial generalized symmetry $v \in \mathcal{V}(M)$ only satisfies

$$\mathcal{L}_v(I_k) \subseteq I_{k+1},$$

so a generalized symmetry v can be *equivalent* (modulo trivials) to one induced by a $v_0 \in \mathcal{V}(M^{(0)})$, without satisfying $\mathcal{L}_v(I_k) \subseteq I_k$. Instead, we have the following.

Definition 4.11 *For a vector field $v \in \mathcal{V}(M)$, the* order *of v, written $o(v)$, is the minimal $k \geq 0$ such that $\mathcal{L}_v(I_0) \subseteq I_{k+1}$.*

With the restriction $o(V) \geq 0$, the orders of equivalent generalized symmetries of $\mathcal{E}$ are equal. A vector field induced by $v_0 \in \mathcal{V}(M^{(0)})$ has order 0. Further properties are:

- $o(v) = k$ if and only if for each $l \geq 0$, $\mathcal{L}_v(I_l) \subseteq I_{l+k+1}$;

- letting $\mathfrak{g}_k = \{v : o(v) \leq k\}$, we have $[\mathfrak{g}_k, \mathfrak{g}_l] \subseteq \mathfrak{g}_{k+l}$.

We now investigate the generalized symmetries of the prolonged contact system on $M = J^\infty(\mathbf{R}^n, \mathbf{R})$. The conclusion will be that the proper generalized symmetries correspond to smooth functions on M; this is analogous to the finite-dimensional contact case, in which we could locally associate to each contact symmetry its generating function, and conversely. Recall that we have a coframing (dx^i, θ_I) for M, satisfying $d\theta_I = -\theta_{Ij} \wedge dx^j$ for all multi-indices I. To describe vector fields on M, we will work with the dual framing $(D_i, \partial/\partial\theta_I)$, which in terms of the usual framing $(\partial/\partial x^i, \partial/\partial p_I)$ is given by

$$D_i = \frac{\partial}{\partial x^i} + \sum_{|I| \geq 0} p_{Ii} \frac{\partial}{\partial p_I},$$

$$\frac{\partial}{\partial \theta_I} = \frac{\partial}{\partial p_I}.$$

The vector fields D_i may be thought of as "total derivative" operators, and applied to a function $g(x^i, z, p_i, \ldots, p_I)$ on some $J^k(\mathbf{R}^n, \mathbf{R})$ give

$$(D_i g)(x^j, z, p_j, \ldots, p_I, p_{Ij}) = \tfrac{\partial g}{\partial x^i} + p_i \tfrac{\partial g}{\partial z} + p_{ij} \tfrac{\partial g}{\partial p_j} + \cdots + p_{Ii} \tfrac{\partial g}{\partial p_I},$$

which will generally be defined only on $J^{k+1}(\mathbf{R}^n, \mathbf{R})$ rather than $J^k(\mathbf{R}^n, \mathbf{R})$. These operators can be composed, and we set

$$D_I = D_{i_1} \circ \cdots \circ D_{i_k}, \qquad I = (i_1, \ldots, i_k).$$

We do this because the proper generalized symmetries of $\mathcal{I} = \{\theta_I : |I| \geq 0\}$ are uniquely represented by vector fields

$$v = g \frac{\partial}{\partial \theta} + g_i \frac{\partial}{\partial \theta^i} + \cdots + g_I \frac{\partial}{\partial \theta_I} + \cdots, \tag{4.40}$$

where $g = v \lrcorner \theta$ and

$$g_I = D_I g, \qquad |I| > 0. \tag{4.41}$$

To see this, first note that any vector field is congruent modulo trivial generalized symmetries to a unique one of the form (4.40). It then follows from a straightforward calculation that a vector field of the form (4.40) is a generalized symmetry of $\mathcal{I}$ if and only if it satisfies (4.41). If one defines $\mathcal{R}_k \subset \mathcal{R}(M)$ to consist of functions pulled back from $J^k(\mathbf{R}^n, \mathbf{R})$, then one can verify that for any proper generalized symmetry $v \in \mathfrak{g}$,

$$o(v) \leq k \quad \Longleftrightarrow \quad g = v \lrcorner \theta \in \mathcal{R}_{k+1}.$$

The general version of Noether's theorem will involve proper generalized symmetries. However, recall that our first-order version requires us to distinguish among the symmetries of an Euler-Lagrange system the symmetries of the original variational problem; only the latter give rise to conservation laws. We therefore have to give the appropriate corresponding notion for proper generalized symmetries.

For this purpose, we introduce the following algebraic apparatus. We filter the differential forms $\Omega^*(M)$ on the infinite prolongation $(M, \mathcal{E})$ of an Euler-Lagrange system $(M^{(0)}, \mathcal{E}^{(0)})$ by letting

$$I^p \Omega^{p+q}(M) = \mathrm{Image}(\underbrace{\mathcal{E} \otimes \cdots \otimes \mathcal{E}}_{p} \otimes \Omega^*(M) \to \Omega^*(M)) \cap \Omega^{p+q}(M). \tag{4.42}$$

We define the associated graded objects

$$\Omega^{p,q}(M) = I^p \Omega^{p+q}(M)/I^{p+1} \Omega^{p+q}(M).$$

Because $\mathcal{E}$ is formally integrable, the exterior derivative d preserves this filtration and its associated graded objects:

$$d : \Omega^{p,q}(M) \to \Omega^{p,q+1}(M).$$

We define the cohomology

$$H_\Lambda^{p,q}(M) = \frac{\mathrm{Ker}(d : \Omega^{p,q}(M) \to \Omega^{p,q+1}(M))}{\mathrm{Im}(d : \Omega^{p,q-1}(M) \to \Omega^{p,q}(M))}.$$

A simple diagram-chase shows that the exterior derivative operator d induces a map $d_1 : H_\Lambda^{p,q}(M) \to H_\Lambda^{p+1,q}(M)$.[8] Now, the Poincaré-Cartan form $\Pi \in \Omega^{n+1}(M^{(0)})$ pulls back to an element $\Pi \in I^2 \Omega^{n+1}(M)$ which is closed and therefore defines a class $[\Pi] \in H_\Lambda^{2,n-1}(M)$.

It follows from the definition that a generalized symmetry of $\mathcal{E}$ preserves the filtration $I^p \Omega^{p+q}(M)$, and therefore acts on the cohomology group $H_\Lambda^{2,n-1}(M)$.

---

[8]Of course, $H_\Lambda^{*,*}(M)$ is the E_1-term of a spectral sequence. Because we will not be using any of the higher terms, however, there is no reason to invoke this machinery. Most of this theory was introduced in [Vin84].

The generalized symmetries appropriate for Noether's theorem are exactly those generalized symmetries v of $\mathcal{E}$ satisfying the additional condition

$$\mathcal{L}_v[\Pi] = 0 \in H_\Lambda^{2,n-1}(M).$$

In other words, v is required to preserve Π modulo (a) forms in $I^3\Omega^{n+1}(M)$, and (b) derivatives of forms in $I^2\Omega^n(M)$. We also need to verify that a *trivial* generalized symmetry v preserves the class $[\Pi]$; this follows from the fact that $v \lrcorner I = 0$, for then $v \lrcorner \Pi \in I^2\Omega^n(M)$, so that

$$\mathcal{L}_v[\Pi] = [d(I^2\Omega^n(M))] = 0.$$

We now have the Lie subalgebra $\mathfrak{g}_{[\Pi]} \subseteq \mathfrak{g}$ of proper generalized symmetries of the variational problem. It is worth noting that this requires only that we have Π defined modulo $I^3\Omega^{n+1}(M)$. A consequence is that even in the most general higher-order, multi-contact case where a canonical Poincaré-Cartan form is not known to exist, there should be a version of Noether's theorem that includes both the first-order multi-contact version discussed in the previous section, and the higher-order scalar version discussed below. However, we will not pursue this.

The other ingredient in Noether's theorem is a space of conservation laws, defined by analogy with previous cases as

$$\mathcal{C}(\mathcal{E}) = H^{n-1}(\Omega^*/\mathcal{E}, \bar{d}) = H_\Lambda^{0,n-1}(M),$$

where the last notation refers to the cohomology just introduced. It is a substantial result (see [BG95a]) that over contractible subsets of $M^{(0)}$ we can use the exterior derivative to identify $\mathcal{C}(\mathcal{E})$ with

$$\bar{\mathcal{C}}(\mathcal{E}) \overset{def}{=} \mathrm{Ker}(d_1 : H_\Lambda^{1,n-1}(M) \to H_\Lambda^{2,n-1}(M)).$$

Now we can identify conservation laws as classes of n-forms, as in the previous case of Noether's theorem, and we will do so without comment in the following.

We define a Noether map $\mathfrak{g}_{[\Pi]} \to \bar{\mathcal{C}}(\mathcal{E})$ as $v \mapsto v \lrcorner \Pi$. To see that this is well-defined, first note that

$$v \lrcorner \Pi \in I^1\Omega^n(M),$$

so $v \lrcorner \Pi$ represents an element of $\Omega^{1,n-1}(M)$, which we shall also denote as $v \lrcorner \Pi$. Furthermore, its exterior derivative is

$$d(v \lrcorner \Pi) = \mathcal{L}_v\Pi, \qquad (4.43)$$

and this lies in $I^2\Omega^{n+1}$, simply because v preserves $\mathcal{E}$ and therefore also the filtration (4.42). Consequently,

$$v \lrcorner \Pi \in \mathrm{Ker}(d : \Omega^{1,n-1}(M) \to \Omega^{1,n}(M)),$$

and we therefore have an element

$$[v \lrcorner \Pi] \in H_\Lambda^{1,n-1}(M).$$

Finally, we need to verify that

$$[v \lrcorner \Pi] \in \mathrm{Ker}(d_1 : H_\Lambda^{1,n-1}(M) \to H_\Lambda^{2,n-1}(M)).$$

This follows from the hypothesis that v preserves not only the Euler-Lagrange system $\mathcal{E}$ and associated filtration (4.42), but also the class $[\Pi]$. Specifically, the image

$$d_1([v \lrcorner \Pi]) \in H_\Lambda^{2,n-1}(M)$$

is represented by the class (see (4.43))

$$[d(v \lrcorner \Pi)] = [\mathcal{L}_v \Pi] = \mathcal{L}_v[\Pi] = 0.$$

This proves that

$$v \mapsto [v \lrcorner \Pi]$$

defines a map between the appropriate spaces.

We can now make the following proposal for a general form of Noether's theorem.

Conjecture 4.1 *Let $(M, \mathcal{E})$ be the infinite prolongation of an Euler-Lagrange system, and assume that the system is non-degenerate and that $H_{dR}^q(M) = 0$ for all $q > 0$. Then the map $v \mapsto [v \lrcorner \Pi]$ induces an isomorphism*

$$\mathfrak{g}_{[\Pi]} \xrightarrow{\sim} \bar{\mathcal{C}}(\mathcal{E}).$$

It is quite possible that this is already essentially proved in [Vin84] or [Olv93], but we have not been able to determine the relationship between their statements and ours. In any case, it would be illuminating to have a proof of the present statement in a spirit similar to that of our Theorem 1.3.

To clarify this, we will describe how it appears in coordinates. First, note that for the classical Lagrangian

$$L(x^i, z, p_i)dx,$$

the Euler-Lagrange equation

$$E(x^i, z, p_i, p_{ij}) = (\textstyle\sum D_j L_{p_j} - L_z)(x^i, z, p_i, p_{ij}) = 0$$

defines a locus $M^{(1)} \subset J^2(\mathbf{R}^n, \mathbf{R})$, and the first prolongation of the Euler-Lagrange system $(J^1(\mathbf{R}^n, \mathbf{R}), \mathcal{E}_L)$ discussed previously is given by the restriction of the second-order contact Pfaffian system on $J^2(\mathbf{R}^n, \mathbf{R})$ to this locus. Higher prolongations are defined by setting

$$M^{(k)} = \{E_I \overset{def}{=} D_I E = 0, \ |I| \le k - 1\} \subset J^{k+1}(\mathbf{R}^n, \mathbf{R})$$

and restricting the $(k+1)$st-order contact system $\mathcal{I}^{(k+1)}$. We will consider generalized symmetries of $(M^{(\infty)}, \mathcal{E}^{(\infty)})$ which arise as restrictions of those generalized symmetries of $(J^{\infty}(\mathbf{R}^n, \mathbf{R}), \mathcal{I})$ which are also tangent to $M^{(\infty)} \subset J^{\infty}(\mathbf{R}^n, \mathbf{R})$. This simplifies matters insofar as we can understand generalized symmetries of $\mathcal{I}$ by their generating functions. The tangency condition is

$$\mathcal{L}_v(E_I)|_{M^{(\infty)}} = 0, \qquad |I| \geq 0. \tag{4.44}$$

This Lie derivative is just the action of a vector field as a derivation on functions. Now, for a generalized symmetry v of the infinite-order contact system, all of the conditions (4.44) follow from just the first one,

$$\mathcal{L}_v(E)|_{M^{(\infty)}} = 0.$$

If we let v have generating function $g = v \lrcorner \theta \in \mathcal{R}(J^{\infty}(\mathbf{R}^n, \mathbf{R}))$, then we can see from (4.40, 4.41) that this condition on g is

$$\sum_{|I| \geq 0} \frac{\partial E}{\partial p_I} D_I g = 0 \text{ on } M^{(\infty)}. \tag{4.45}$$

We are again using $p = z$ for convenience. For instance, $E = \sum p_{ii} - f(z)$ defines the Poisson equation $\Delta z = f(z)$, and the preceding condition is

$$\sum_i D_i^2 g - f'(z)g = 0 \text{ on } M^{(\infty)}. \tag{4.46}$$

We now consider the Noether map for $M^{(\infty)} \subset J^{\infty}(\mathbf{R}^n, \mathbf{R})$. We write the Poincaré-Cartan form pulled back to $M^{(\infty)}$ using coframes adapted to this infinite prolongation, starting with

$$dL_{p_i} = D_j(L_{p_i})dx^j + L_{p_i z}\theta + L_{p_i p_j}\theta_j,$$

and then the Poincaré-Cartan form on $J^{\infty}(\mathbf{R}^n, \mathbf{R})$ is

$$\begin{aligned} \Pi &= d(L\,dx + \theta \wedge L_{p_i}\,dx_{(i)}) \\ &= \theta \wedge \left((-D_i(L_{p_i}) + L_z)dx - \theta_j \wedge L_{p_i p_j}\,dx_{(i)}\right). \end{aligned}$$

Restriction to $M^{(\infty)} \subset J^{\infty}(\mathbf{R}^n, \mathbf{R})$ kills the first term, and we have

$$\Pi = -L_{p_i p_j}\theta \wedge \theta_j \wedge dx_{(i)}.$$

We then apply a vector field v_g with generating function g, and obtain

$$v_g \lrcorner \Pi = -gL_{p_i p_j}\theta_j \wedge dx_{(i)} + (D_j g)L_{p_i p_j}\theta \wedge dx_{(i)}.$$

This will be the "differentiated form" of a conservation law precisely if $\mathcal{L}_{v_g}[\Pi] = 0$, that is, if

$$d(v_g \lrcorner \Pi) \equiv 0 \pmod{I^3 \Omega^{n+1}(M) + dI^2 \Omega^n(M)}.$$

Concerning generalized symmetries of a PDE, note that in the condition (4.45) for $g = g(x^i, p, p_i, \ldots, p_I) \in \mathcal{R}_k = C^\infty(J^k(\mathbf{R}^n, \mathbf{R}))$, the variables p_I with $|I| > k$ appear only polynomially upon taking the total derivatives $D_J g$. In other words, the condition on g is polynomial in the variables p_I for $|I| > k$. Equating coefficients of these polynomials gives a PDE system to be satisfied by a generalized symmetry of an Euler-Lagrange equation. With some effort, one can analyze the situation for our Poisson equation $\Delta z = f(z)$ and find the following.

Proposition 4.6 *If $n \geq 3$, then a solution $g = g(x^i, p, p_i, \ldots, p_I)$ of order k to (4.46) is equal on $M^{(\infty)}$ to a function that is linear in the variables p_J with $|J| \geq k - 2$. If in addition $f''(z) \neq 0$, so that the Poisson equation is non-linear, then every solution's restriction to $M^{(\infty)}$ is the pullback of a function on $M^{(0)} \subset J^2(\mathbf{R}^n, \mathbf{R})$, which generates a classical symmetry of the equation.*

In other words, a non-linear Poisson equation in $n \geq 3$ independent variables has no non-classical generalized symmetries, and consequently no higher-order conservation laws.

Proof. Because the notation involved here becomes rather tedious, we will sketch the proof and leave it to the reader to verify the calculations. We previously hinted at the main idea: the condition (4.46) on a generating function $g = g(x^i, p, \ldots, p_I)$, $|I| = k$, is polynomial in the highest order variables with coefficients depending on partial derivatives of g. To isolate these terms, we filter the functions on $M^{(\infty)}$ by letting $\overline{\mathcal{R}_l}$ be the image of $\mathcal{R}_l$ under restriction to $M^{(\infty)}$; in other words, $\overline{\mathcal{R}_l}$ consists of functions which can be expressed as functions of x^i, p_J, $|J| \leq l$, *after* substituting the defining relations of $M^{(\infty)}$,

$$p_{Jii} = \frac{d^{|J|} f}{dx^J}.$$

To calculate in $\overline{\mathcal{R}_l}$ we will need to define variables q_J to be the harmonic parts of p_J; that is,

$$
\begin{aligned}
q_i &= p_i, \\
q_{ij} &= p_{ij} - \tfrac{1}{n}\delta_{ij} p_{ll} \\
&= p_{ij} - \tfrac{1}{n}\delta_{ij} f(p), \\
q_{ijk} &= p_{ijk} - \tfrac{1}{n+2}(\delta_{ij} p_{kll} + \delta_{jk} p_{ill} + \delta_{ki} p_{jll}) \\
&= p_{ijk} - \tfrac{1}{n+2}(\delta_{ij} p_k + \delta_{jk} p_i + \delta_{ki} p_j) f'(p), \quad \&\text{c.}
\end{aligned}
$$

These, along with x_i and p, give coordinates on $M^{(\infty)}$. In addition to working modulo various $\overline{\mathcal{R}_l}$ to isolate terms with higher-order derivatives, we will also at times work modulo functions that are *linear* in the q_I. In what follows, we use the following index conventions: $p^{(l)} = (p, p_j, \ldots, p_J)$ (with $|J| = l$) denotes the derivative variables up to order l, and the multi-indices I, K, A, satisfy $|I| = k$, $|K| = k - 1$, $|A| = k - 2$.

Now, starting with $g = g(x^j, p^{(k)}) \in \mathcal{R}_k$, we note that

$$0 = \sum_i D_i^2 g - f' g \in \overline{\mathcal{R}_{k+1}};$$

that is, the possible order-$(k+2)$ term resulting from two differentiations of g already drops to order k when restricted to the equation manifold. We consider this expression modulo $\overline{\mathcal{R}_k}$,[9] and obtain a quadratic polynomial in q_{Ij} with coefficients in $\overline{\mathcal{R}_k}$. We consider only the quadratic terms of this polynomial, which are

$$0 \equiv \sum_{\substack{|I|,|I'|=k \\ 1 \le i \le n}} \frac{\partial^2 g}{\partial p_I \partial p_{I'}} q_{Ii} q_{I'i}. \tag{4.47}$$

To draw conclusions about $\frac{\partial^2 g}{\partial p_I \partial p_{I'}}$ from this, we need the following fundamental lemma, in which the difference between the cases $n = 2$ and $n \ge 3$ appears:

> If $\mathbf{H}_l \subset Sym^l(\mathbf{R}^n)^*$ denotes the space of degree-l homogeneous harmonic polynomials on $\mathbf{R}^n$, $n \ge 3$, then the $O(n)$-equivariant contraction map $\mathbf{H}_{l+1} \otimes \mathbf{H}_{m+1} \to \mathbf{H}_l \otimes \mathbf{H}_m$ given by
>
> $$X_C z^C \otimes Y_Q z^Q \mapsto \sum_i X_{Bi} z^B \otimes Y_{Pi} z^P$$
>
> is surjective; here, $|C| - 1 = |B| = l$, $|Q| - 1 = |P| = m$.

We will apply this in situations where a given $g^{BP} \in Sym^l(\mathbf{R}^n) \otimes Sym^m(\mathbf{R}^n)$ is known to annihilate all $X_{Bi} Y_{Pi}$ with X_C, Y_Q harmonic, for then we have g^{BP} orthogonal to $\mathbf{H}_l \otimes \mathbf{H}_m \subset Sym^l(\mathbf{R}^n)^* \otimes Sym^m(\mathbf{R}^n)^*$. In particular, from (4.47) we have

$$\frac{\partial^2 g}{\partial p_I \partial p_J} q_I q_J \equiv 0.$$

This means that the restriction of the function g to the hyperplanes $p_{Aii} = \frac{d^{|A|} f}{dx^A}$ is linear in the highest p_I; in other words, we can write

$$g(x^i, p^{(k)}) = h(x, p^{(k-1)}) + h^I(x, p^{(k-1)}) p_I$$

for some $h^I \in \mathcal{R}_{k-1}$. We can further assume that all $h^{Aii} = 0$, where $|A| = k-2$. This completes the first step.

The second step is to simplify the functions $h^I(x, p^{(k-1)})$, substituting our new form of g into the condition (4.46). Again, the "highest" terms appear modulo $\overline{\mathcal{R}_k}$, and are

$$0 \equiv 2 \frac{\partial h^I}{\partial p_K} q_{Ki} q_{Ii} + 2 \frac{\partial h^I}{\partial p_A} q_{Ai} q_{Ii};$$

---

[9]In this context, "modulo" refers to quotients of vector spaces by subspaces, not of rings by ideals, as in exterior algebra.

here we recall our index convention $|A| = k-2$, $|K| = k-1$, $|I| = k$. Both terms must vanish separately, and for the first, our lemma on harmonic polynomials gives that $[h]^{IK} \overset{def}{=} \frac{\partial h^I}{\partial p_K}$ is orthogonal to harmonics; but then our normalization hypothesis $h^{Aii} = 0$ gives that $[h]^{IK} = 0$. We conclude that

$$g(x^i, p^{(k)}) = h(x, p^{(k-1)}) + h^I(x, p^{(k-2)})p_I, \qquad |I| = k.$$

For the second term, our lemma gives similarly that $\frac{\partial h^I}{\partial p_A} = 0$, so we have

$$g(x^i, p^{(k)}) = h(x, p^{(k-1)}) + h^I(x, p^{(k-3)})p_I.$$

This completes the second step.

For the third step, we again substitute the latest form of g into the condition (4.46), and now work modulo $\mathcal{R}_{k-1}$. The only term non-linear in the p_I is

$$0 \equiv \frac{\partial^2 h}{\partial p_K \partial p_{K'}} p_{Ki} p_{K'i},$$

and as before, the lemma implies that h is linear in p_K. Now we have

$$g(x^i, p^{(k)}) = h(x, p^{(k-2)}) + h^K(x, p^{(k-2)})p_K + h^I(x, p^{(k-3)})p_I.$$

Again working modulo $\mathcal{R}_{k-1}$, the only term that is non-linear in (p_K, p_I) is $\frac{\partial h^K}{\partial p_A} p_{Ai} p_{Ki}$, so as before, we can assume $\frac{\partial h^K}{\partial p_A} = 0$ and write

$$g(x^i, p^{(k)}) = h(x, p^{(k-2)}) + h^K(x, p^{(k-3)})p_K + h^I(x, p^{(k-3)})p_I.$$

The final step is similar, and gives that h is linear in p_A. This yields

$$g(x^i, p^{(k)}) = h + h^A p_A + h^K p_K + h^I p_I, \tag{4.48}$$

where each of h, h^A, h^K, h^I is a function of $(x^i, p^{(k-3)})$. This is the first statement of the proposition.

To derive the second statement, we use the form (4.48) in the condition (4.46) modulo $\overline{\mathcal{R}}_{k-2}$, in which the only term *non-linear* in (p_K, p_I) is

$$\begin{aligned} 0 &\equiv h^I(x^i, p^{(k-3)})\frac{d^{|I|}f}{dx^I} \\ &\equiv h^I(x^i, p^{(k-3)})f''(p)\sum_{i \in I} p_{I\setminus i}p_i. \end{aligned}$$

In particular, if $f''(p) \neq 0$, then we must have $h^I = 0$. Therefore $g \in \mathcal{R}_k$ actually lies in $\mathcal{R}_{k-1}$, and we can induct downward on k, eventually proving that $g \in \mathcal{R}_1$, as desired. $\qquad\square$

We mention two situations which contrast sharply with that of the non-linear Poisson equation in dimension $n \geq 3$. First, in the case of a *linear* Poisson equation $\Delta z = f(z)$, $f''(z) = 0$ (still in $n \geq 3$ dimensions), one can extend the preceding argument to show that a generating function for a conservation law is linear in all of the derivative variables p_I. In particular, the infinite collection of conservation laws for the Laplace equation $\Delta z = 0$ can be determined in this manner; it is interesting to see how all of these disappear upon the addition of a non-linear term to the equation.

Second, in dimension $n = 2$, there are well-known non-linear Poisson equations $\Delta u = \sinh u$ and $\Delta u = e^u$ having infinitely many higher-order conservation laws, but we will not discuss these.

4.3.3   The $K = -1$ Surface System

In §1.4, we constructed Monge-Ampere systems on the contact manifold M^5 of oriented tangent planes to Euclidean space $\mathbf{E}^3$, whose integral manifolds corresponded to linear Weingarten surfaces. We briefly recall this setup for the case of surfaces with Gauss curvature $K = -1$. Our index ranges are now $1 \leq a, b, c \leq 3$, $1 \leq i, j, k \leq 2$.

Let $\mathcal{F} \to \mathbf{E}^3$ be the Euclidean frame bundle, with its global coframing ω^a, $\omega_b^a = -\omega_a^b$ satisfying the structure equations

$$d\omega^a = -\omega_b^a \wedge \omega^b, \quad d\omega_b^a = -\omega_c^a \wedge \omega_b^c.$$

We set $\theta = \omega^3$, $\pi_i = \omega_i^3$, and then $\theta \in \Omega^1(\mathcal{F})$ is the pullback of a global contact form on $M = G_2(T\mathbf{E}^3)$. The forms that are semibasic over M are generated by $\theta, \omega^i, \pi_i \in \Omega^1(\mathcal{F})$.

We define the 2-forms on $\mathcal{F}$

$$\begin{aligned}
\Theta &= d\theta = -\pi_1 \wedge \omega^1 - \pi_2 \wedge \omega^2, \\
\Psi &= \pi_1 \wedge \pi_2 + \omega^1 \wedge \omega^2,
\end{aligned}$$

which are pullbacks of uniquely determined forms on M. On a transverse integral element $E^2 \subset T_m M$ of the contact system $\mathcal{I} = \{\theta, \Theta\}$, on which $\omega^1 \wedge \omega^2 \neq 0$, there are relations

$$\pi_i = h_{ij}\omega^j, \quad h_{ij} = h_{ji}.$$

In this case,

$$\pi_1 \wedge \pi_2 = K\omega^1 \wedge \omega^2,$$

where $K = h_{11}h_{22} - h_{12}h_{21}$ is the Gauss curvature of any surface $N^2 \hookrightarrow \mathbf{E}^3$ whose 1-jet graph in M^5 is tangent to $E \subset T_m M$. Therefore, transverse integral manifolds of the EDS

$$\mathcal{E} = \{\theta, \Theta, \Psi\} \tag{4.49}$$

correspond locally to surfaces in $\mathbf{E}^3$ with constant Gauss curvature $K = -1$.

The EDS $(M, \mathcal{E})$ is an example of a hyperbolic Monge-Ampere system; this notion appeared in §2.1, where we used it to specify a branch of the equivalence problem for Poincaré-Cartan forms on contact 5-manifolds. The defining property of a hyperbolic Monge-Ampere system $\mathcal{E} = \{\theta, \Theta, \Psi\}$ is that modulo the algebraic ideal $\{\theta\}$, $\mathcal{E}$ contains two distinct (modulo scaling) *decomposable* 2-forms; that is, one can find two non-trivial linear combinations of the form

$$\begin{aligned}
\lambda_1 \Theta + \mu_1 \Psi &= \alpha_1 \wedge \beta_1, \\
\lambda_2 \Theta + \mu_2 \Psi &= \alpha_2 \wedge \beta_2.
\end{aligned}$$

This exhibits two rank-2 Pfaffian systems $I_i = \{\alpha_i, \beta_i\}$, called the *characteristic systems* of $\mathcal{E}$, which are easily seen to be independent of choices (except for which one is I_1 and which one is I_2). The relationship between the geometry of the characteristic systems and that of the original hyperbolic Monge-Ampere system is very rich (see [BGH95]). Of particular interest are those hyperbolic systems whose characteristic systems each contain a non-trivial conservation law. We will show that this holds for the $K = -1$ system introduced above, but only after one prolongation. In other words, for the prolonged system $\mathcal{E}^{(1)}$, there is also a notion of characteristic systems $I_i^{(1)}$ which restrict to any integral surface as the original I_i, and each of these $I_i^{(1)}$ contains a non-trivial conservation law for $\mathcal{E}^{(1)}$.

Returning to the discussion of integral elements of $\mathcal{E} = \{\theta, \Theta, \Psi\}$, note that for any integral element $E \subset T_{(p,H)}M$, given by equations

$$\pi_i - h_{ij}\omega^j = 0,$$

there is a unique frame $(p, (e_1, e_2, e_1 \times e_2)) \in \mathcal{F}$ over $(p, e_1 \wedge e_2) \in M$ for which the second fundamental form is normalized as

$$h_{11} = a > 0, \quad h_{22} = -\tfrac{1}{a}, \quad h_{12} = h_{21} = 0.$$

The tangent lines in $\mathbf{E}^3$ spanned by these e_1, e_2 are the *principal directions* at p of any surface whose 1-jet graph is tangent to E; they define an orthonormal frame in which the second fundamental form is diagonal. The number $a > 0$ is determined by the plane $E \subset TM$, so to study integral elements of $(M, \mathcal{E})$, and in particular to calculate on its first prolongation, we introduce

$$\mathcal{F}^{(1)} = \mathcal{F} \times \mathbf{R}^*,$$

where $\mathbf{R}^*$ has the coordinate $a > 0$. There is a projection $\mathcal{F}^{(1)} \to M^{(1)}$, mapping

$$(p, e, a) \mapsto (p, e_1 \wedge e_2, \{\pi_1 - a\omega^1, \pi_2 + \tfrac{1}{a}\omega^2\}^\perp).$$

We define on $\mathcal{F}^{(1)}$ the forms

$$\begin{aligned}
\theta_1 &= \pi_1 - a\omega^1, \\
\theta_2 &= \pi_2 + \tfrac{1}{a}\omega^2,
\end{aligned}$$

which are semibasic for $\mathcal{F}^{(1)} \to M^{(1)}$. The first prolongation of the system $\mathcal{E}$ on M is a Pfaffian system on $M^{(1)}$, which pulls back to $\mathcal{F}^{(1)}$ as

$$\mathcal{E}^{(1)} = \{\theta, \theta_1, \theta_2, d\theta_1, d\theta_2\}.$$

We have structure equations

$$\left.\begin{array}{l} d\theta \;\equiv\; -\theta_1 \wedge \omega^1 - \theta_2 \wedge \omega^2 \equiv 0 \\ d\theta_1 \;\equiv\; -da \wedge \omega^1 + \frac{1+a^2}{a}\omega_2^1 \wedge \omega^2 \\ d\theta_2 \;\equiv\; \frac{1}{a^2}da \wedge \omega^2 + \frac{1+a^2}{a}\omega_2^1 \wedge \omega^1 \end{array}\right\} \quad (\mathrm{mod}\ \{\theta, \theta_1, \theta_2\}),$$

and in particular, we have the decomposable linear combinations

$$\begin{array}{l} -d\theta_1 - a\, d\theta_2 = (da - (1 + a^2)\omega_2^1) \wedge (\omega^1 + \tfrac{1}{a}\omega^2), \\ -d\theta_1 + a\, d\theta_2 = (da + (1 + a^2)\omega_2^1) \wedge (\omega^1 - \tfrac{1}{a}\omega^2). \end{array} \tag{4.50}$$

The EDS $\mathcal{E}^{(1)}$ is algebraically generated by $\theta, \theta_1, \theta_2$, and these two decomposable 2-forms. The characteristic systems are by definition differentially generated by

$$\begin{array}{l} I_1^{(1)} = \{\theta,\ \theta_1,\ \theta_2,\ da - (1 + a^2)\omega_2^1,\ \omega^1 + \tfrac{1}{a}\omega^2\}, \\ I_2^{(1)} = \{\theta,\ \theta_1,\ \theta_2,\ da + (1 + a^2)\omega_2^1,\ \omega^1 - \tfrac{1}{a}\omega^2\}. \end{array} \tag{4.51}$$

Now, the "universal" second fundamental form can be factored as

$$II = a(\omega^1)^2 - \tfrac{1}{a}(\omega^2)^2 = a(\omega^1 + \tfrac{1}{a}\omega^2)(\omega^1 - \tfrac{1}{a}\omega^2). \tag{4.52}$$

These linear factors, restricted any $K = -1$ surface, define the *asymptotic curves* of that surface, so by comparing (4.51) and (4.52) we find that:

> On a $K = -1$ *surface, the integral curves of the characteristic systems are the asymptotic curves.*

Now we look for Euclidean-invariant conservation laws in each $I_i^{(1)}$. Instead of using Noether's theorem, we work directly. We start by setting

$$\varphi_1 = f(a)(\omega^1 + \tfrac{1}{a}\omega^2) \in I_1^{(1)},$$

and seek conditions on $f(a)$ to have $d\varphi_1 \in \mathcal{E}^{(1)}$. A short computation using the structure equations gives

$$d\varphi_1 \equiv (f'(a)(1 + a^2) - f(a)a)\omega_2^1 \wedge (\omega^1 + \tfrac{1}{a}\omega^2) \quad (\mathrm{mod}\ \mathcal{E}^{(1)}),$$

so the condition for φ_1 to be a conserved 1-form is

$$\frac{f'(a)}{f(a)} = \frac{a}{1 + a^2}.$$

A solution is

$$\varphi_1 = \tfrac{1}{2}\sqrt{1 + a^2}(\omega^1 + \tfrac{1}{a}\omega^2);$$

the choice of multiplicative constant $\tfrac{1}{2}$ will simplify later computations. A similar computation, seeking an appropriate multiple of $\omega^1 - \tfrac{1}{a}\omega^2$, yields the conserved 1-form

$$\varphi_2 = \tfrac{1}{2}\sqrt{1 + a^2}(\omega^1 - \tfrac{1}{a}\omega^2).$$

On any simply connected integral surface of the $K = -1$ system, there are coordinate functions s, t such that

$$\varphi_1 = ds, \qquad \varphi_2 = dt.$$

If we work in these coordinates, and in particular use the non-orthonormal coframing (φ_1, φ_2) then we can write

$$\omega^1 = \left(\frac{1}{\sqrt{1+a^2}}\right)(\varphi_1 + \varphi_2), \tag{4.53}$$

$$\omega^2 = \left(\frac{a}{\sqrt{1+a^2}}\right)(\varphi_1 - \varphi_2), \tag{4.54}$$

$$I = (\omega^1)^2 + (\omega^2)^2 = (\varphi_1)^2 + 2\left(\frac{1-a^2}{1+a^2}\right)\varphi_1\varphi_2 + (\varphi_2)^2, \tag{4.55}$$

$$II = \left(\frac{4a}{1+a^2}\right)\varphi_1\varphi_2. \tag{4.56}$$

These expressions suggest that we define

$$a = \tan z,$$

where $a > 0$ means that we can smoothly choose $z = \tan^{-1} a \in (0, \pi/2)$. Note that $2z$ is the angle measure between the asymptotic directions $\varphi_1^\perp, \varphi_2^\perp$, and that

$$\omega^1 = (\cos z)(\varphi_1 + \varphi_2), \quad \omega^2 = (\sin z)(\varphi_1 - \varphi_2). \tag{4.57}$$

The following is fundamental in the study of $K = -1$ surfaces.

Proposition 4.7 *On an immersed surface in* $\mathbf{E}^3$ *with constant Gauss curvature* $K = -1$, *the associated function* z, *expressed in terms of asymptotic coordinates* s, t, *satisfies the* sine-Gordon equation

$$z_{st} = \tfrac{1}{2}\sin(2z). \tag{4.58}$$

One can prove this by a direct computation, but we will instead highlight certain general EDS constructions which relate the $K = -1$ differential system to a hyperbolic Monge-Ampere system associated to the sine-Gordon equation. One of these is the notion of an *integrable extension* of an exterior differential system, which we have not yet encountered. This is a device that handles a forseeable difficulty; namely, the sine-Gordon equation is expressed in terms of the variables s and t, but for the $K = -1$ system, these are primitives of a conservation law, defined on integral manifolds of the system only up to addition of integration constants. One can think of an integrable extension as a device for appending the primitives of conserved 1-forms. More precisely, an integrable extention of an EDS $(M, \mathcal{E})$ is given by a submersion $M' \xrightarrow{\pi} M$, with a differential ideal $\mathcal{E}'$ on M' generated *algebraically* by $\pi^*\mathcal{E}$ and some 1-forms on M'. In this case, the preimage in M' of an integral manifold of $\mathcal{E}$ is foliated by integral manifolds of $\mathcal{E}'$. For example, if a 1-form $\varphi \in \Omega^1(M)$ is a conservation law for $\mathcal{E}$, then one can take $M' = M \times \mathbf{R}$, and let $\mathcal{E}' \subset \Omega^*(M')$ be generated by $\mathcal{E}$ and $\tilde{\varphi} = \varphi - ds$, where s is a fiber coordinate on $\mathbf{R}$. Then the preimage in M' of any integral manifold of $(M, \mathcal{E})$ is foliated by a 1-parameter family of integral manifolds of $(M', \mathcal{E}')$, where the parameter corresponds to a choice of integration constant for φ.[10]

Proof. Because z is defined only on $M^{(1)}$, it is clear that we will have to prolong once more to study z_{st}. (Twice is unnecessary, because of the Monge-Ampere form of (4.58).) From (4.50), we see that integral elements with $\varphi_1 \wedge \varphi_2 \neq 0$ satisfy

$$dz - \omega_2^1 = 2p\varphi_1, \quad dz + \omega_2^1 = 2q\varphi_2,$$

for some p, q. These p, q can be taken as fiber coordinates on the second prolongation

$$M^{(2)} = M^{(1)} \times \mathbf{R}^2.$$

Let

$$\theta_3 = dz - p\varphi_1 - q\varphi_2,$$
$$\theta_4 = \omega_2^1 + p\varphi_1 - q\varphi_2,$$

and then the prolonged differential system is

$$\mathcal{E}^{(2)} = \{\theta, \theta_1, \ldots, \theta_4, d\theta_3, d\theta_4\}.$$

Integral manifolds for the original $K = -1$ system correspond to integral manifolds of $\mathcal{E}^{(2)}$; in particular, on such an integral manifold $f^{(2)} : N \hookrightarrow M^{(2)}$ we have

$$0 = d\theta_3|_{N^{(2)}} = -dp \wedge \varphi_1 - dq \wedge \varphi_2$$

and

$$
\begin{aligned}
0 &= d\theta_4|_{N^{(2)}} \\
&= K\omega^1 \wedge \omega^2 + dp \wedge \varphi_1 - dq \wedge \varphi_2 \\
&= (-1)(\cos z)(\varphi_1 + \varphi_2) \wedge (\sin z)(\varphi_1 - \varphi_2) + dp \wedge \varphi_1 - dq \wedge \varphi_2 \\
&= \sin(2z)\varphi_1 \wedge \varphi_2 + dp \wedge \varphi_1 - dq \wedge \varphi_2.
\end{aligned}
$$

Now we define the integrable extension

$$M^{(2)\prime} = M^{(2)} \times \mathbf{R}^2,$$

where $\mathbf{R}^2$ has coordinates s, t, and on $M^{(2)\prime}$ we define the EDS $\mathcal{E}^{(2)\prime}$ to be generated by $\mathcal{E}^{(2)}$, along with the 1-forms $\varphi_1 - ds$, $\varphi_2 - dt$. An integral manifold $f^{(2)} : N \hookrightarrow M^{(2)}$ of $\mathcal{E}^{(2)}$ gives a 2-parameter family of integral manifolds $f^{(2)}_{s_0, t_0}$:

---

[10]For more information about integrable extensions, see §6 of [BG95b].

$N \hookrightarrow M^{(2)\prime}$ of $\mathcal{E}^{(2)\prime}$. On any of these, the functions s, t will be local coordinates, and we will have

$$0 = dz - p\,ds - q\,dt,$$
$$0 = -dp \wedge ds - dq \wedge dt,$$
$$0 = \sin(2z)ds \wedge dt - ds \wedge dp - dq \wedge dt.$$

These three clearly imply that $z(s, t)$ satisfies (4.58). $\qquad\square$

Note that we can start from the other side, defining the differential system for the sine-Gordon equation as

$$\mathcal{E}_{sG} = \{dz - p\,ds - q\,dt, -dp \wedge ds - dq \wedge dt, ds \wedge dp + dq \wedge dt - \sin(2z)ds \wedge dt\},$$

which is a Monge-Ampere system on the contact manifold $J^1(\mathbf{R}^2, \mathbf{R})$. One can form a "non-abelian" integrable extension

$$\mathcal{P} = J^1(\mathbf{R}^2, \mathbf{R}) \times \mathcal{F}$$

of $(J^1(\mathbf{R}^2, \mathbf{R}), \mathcal{E}_{sG})$ by taking

$$\mathcal{E}_\mathcal{P} = \mathcal{E}_{sG} + \left\{ \begin{array}{l} \omega^1 - (\cos z)(ds + dt), \\ \omega^2 - (\sin z)(ds + dt), \\ \omega^3, \\ \omega_2^1 + p\,ds - q\,dt, \\ \omega_1^3 - (\sin z)(ds + dt), \\ \omega_2^3 + (\cos z)(ds - dt). \end{array} \right\}.$$

This system is differentially closed, as one can see by using the structure equations for $\mathcal{F}$ and assuming that z satisfies the sine-Gordon equation. Though the following diagram is complicated, it sums up the whole story:

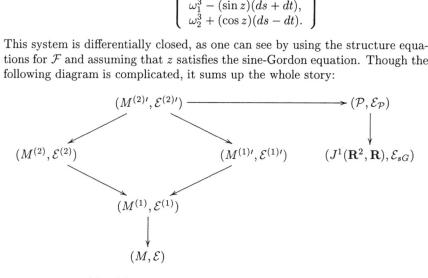

The system $(M^{(1)\prime}, \mathcal{E}^{(1)\prime})$ was not introduced in the proof; it is the integrable extension of $(M^{(1)}, \mathcal{E}^{(1)})$ formed by adjoining primitives s, t for the conserved 1-forms φ_1, φ_2, and its prolongation turns out to be $(M^{(2)\prime}, \mathcal{E}^{(2)\prime})$. In other words, starting on $M^{(1)}$, one can first prolong and then adjoin primitives, or vice versa. The main point of this diagram is:

The identification $(M^{(2)\prime}, \mathcal{E}^{(2)\prime}) \longrightarrow (\mathcal{P}, \mathcal{E}_\mathcal{P})$ is an isomorphism of exterior differential systems. In other words, modulo prolongations and integrable extensions, the $K = -1$ system and the sine-Gordon system are equivalent.

Note that while conservation laws are preserved under prolongation, there is an additional subtlety for integrable extensions. In this case, only those conservation laws for the sine-Gordon system that are invariant under s, t-translation give conservation laws for the $K = -1$ system. Conversely, only those conservation laws for the $K = -1$ system that are invariant under Euclidean motions (i.e., translations in $\mathcal{F}$) give conservation laws for the sine-Gordon system. There is even a difficulty involving trivial conservation laws; namely, the non-trivial, Euclidean-invariant conservation laws φ_1, φ_2 for the $K = -1$ system induce the trivial conservation laws ds, dt for the sine-Gordon system.

However, we do have two conservation laws for sine-Gordon, obtained via Noether's theorem applied to s, t-translations and the Lagrangian

$$\Lambda_{sG} = (pq - \tfrac{1}{2}(\cos(2z) - 1))ds \wedge dt;$$

they are

$$\psi_1 = \tfrac{1}{2}p^2 ds - \tfrac{1}{4}(\cos(2z) - 1)dt,$$
$$\psi_2 = \tfrac{1}{2}q^2 dt + \tfrac{1}{4}(\cos(2z) - 1)ds.$$

The corresponding conserved 1-forms on $(M^{(2)}, \mathcal{E}^{(2)})$ are

$$\varphi_3 = \tfrac{1}{2}p^2 \varphi_1 - \tfrac{1}{4}(\cos(2z) - 1)\varphi_2,$$
$$\varphi_4 = \tfrac{1}{2}q^2 \varphi_2 + \tfrac{1}{4}(\cos(2z) - 1)\varphi_1.$$

In the next section, we will introduce a Bäcklund transformation for the sine-Gordon equation, which can be used to generate an infinite double-sequence ψ_{2k-1}, ψ_{2k} of conservation laws (see [AI79]). These in turn give an infinite double-sequence $\varphi_{2k+1}, \varphi_{2k+1}$ of independent conservation laws for $K = -1$, extending the two pairs that we already have. Although we will not discuss these, it is worth pointing out that the generalized symmetries on $(M^{(\infty)}, \mathcal{E}^{(\infty)})$ to which they correspond under Noether's theorem are *not* induced by symmetries at any finite prolongation. For this reason, they are called *hidden symmetries*.

Finally, we mention the following non-existence result, which is similar to the result in Proposition 4.6 for higher-dimensional non-linear Poisson equations.

Proposition 4.8 *In dimension $n \geq 3$, there are no second-order Euclidean-invariant conservation laws for the linear Weingarten system for hypersurfaces in $\mathbf{E}^{n+1}$ with Gauss curvature $K = -1$.*

Proof. In contrast to our proof of the analogous statement for non-linear Poisson equations, we give here a direct argument not appealing to generating functions. We work on the product

$$\mathcal{F}^{(1)} = \mathcal{F} \times \{(a_1, \ldots, a_n) \in \mathbf{R}^n : \prod a_i = -1\},$$

where $\mathcal{F}$ is the Euclidean frame bundle for $\mathbf{E}^{n+1}$, and the other factor parameterizes eigenvalues of admissible second fundamental forms. We use the usual structure equations, but without the sum convention:

$$d\omega_i = -\sum_j \omega_{ij} \wedge \omega_j + \pi_i \wedge \theta,$$

$$d\theta = -\sum_j \pi_j \wedge \omega_j$$

$$d\omega_{ij} = -\sum_k \omega_{ik} \wedge \omega_{kj} + \pi_i \wedge \pi_j,$$

$$d\pi_i = -\sum_j \pi_j \wedge \omega_{ji}.$$

There is a Pfaffian system $\mathcal{I}$ on $\mathcal{F}^{(1)}$ whose transverse n-dimensional integral manifolds correspond to $K = -1$ hypersurfaces; it is differentially generated by θ and the n 1-forms

$$\theta_i = \pi_i - a_i\omega_i.$$

A conservation law for this system is an $(n-1)$-form on $\mathcal{F}^{(1)}$ whose exterior derivative vanishes on any integral manifold of $\mathcal{I}$. A conservation law is invariant under Euclidean motions if its restriction has the form

$$\varphi = \sum f_i(a_1, \ldots, a_n)\omega_{(i)}.$$

What we will show is that for $n \geq 3$, such a form cannot be closed modulo $\mathcal{I}$ unless it equals 0.

First, we calculate using the structure equations that

$$d\varphi \equiv \sum_i (df_i - \sum_j f_j\omega_{ij}) \wedge \omega_{(i)} \quad (\text{mod } \mathcal{I}). \tag{4.59}$$

To proceed further, we want an expression for $\omega_{ij} \wedge \omega_{(i)}$, which we obtain first by computing

$$0 \equiv d\theta_i \quad (\text{mod } \mathcal{I})$$

$$\equiv -\sum_j \pi_k \wedge \omega_{ki} - da_i \wedge \omega_i + a_i \sum_k \omega_{ik} \wedge \omega_k$$

$$\equiv -da_i \wedge \omega_i + \sum_k (a_i - a_k)\omega_{ik} \wedge \omega_k.$$

and then by multiplying the last result by $\omega_{(ij)}$:

$$0 \equiv da_i \wedge \omega_{(j)} + (a_i - a_j)\omega_{ij} \wedge \omega_{(i)} \quad (\text{mod } \mathcal{I}).$$

Using this in (4.59), we obtain

$$d\varphi \equiv \sum_i \left(df_i + \sum_j \frac{f_i}{a_j - a_i} da_j \right) \wedge \omega_{(i)} \quad (\text{mod } \mathcal{I}).$$

So φ is a conservation law only if for each i,

$$df_i = f_i \sum_j \frac{da_j}{a_i - a_j}.$$

Keep in mind that we are requiring this equation to hold on the locus $\mathcal{F}^{(1)}$ where $\Pi a_i = -1$. Wherever at least one $f_i(a_1, \ldots, a_n)$ is non-zero, we must have

$$0 = d\left(\sum_j \frac{da_j}{a_i - a_j}\right) = -\sum_j \frac{da_i \wedge da_j}{(a_i - a_j)^2}.$$

However, when $n \geq 3$, the summands in the expression are linearly independent 2-forms on $\mathcal{F}^{(1)}$. □

It is worth noting that when $n = 2$, this 2-form *does* vanish, and we can solve for $f_1(a_1, a_2)$, $f_2(a_1, a_2)$ to obtain the conservation laws φ_1, φ_2 discussed earlier. The same elementary method can be used to analyze second-order conservation laws for more general Weingarten equations; in this way, one can obtain a full classification of those few Wiengarten equations possessing higher-order conservation laws.

4.3.4   Two Bäcklund Transformations

We have seen a relationship between the $K = -1$ surface system, and the sine-Gordon equation

$$z_{xy} = \tfrac{1}{2}\sin(2z). \tag{4.60}$$

Namely, the half-angle measure between the asymptotic directions on a $K = -1$ surface, when expressed in asymptotic coordinates, satisfies the sine-Gordon equation. We have also interpreted this relationship in terms of important EDS constructions. In this section, we will explain how this relationship connects the *Bäcklund transformations* associated to each of these systems.

There are many definitions of Bäcklund transformation in the literature, and instead of trying to give an all-encompassing definition, we will restrict attention to Monge-Ampere systems

$$\mathcal{E} = \{\theta, \Theta, \Psi\},$$

where θ is a contact form on a manifold (M^5, I), and $\Theta, \Psi \in \Omega^2(M)$ are linearly independent modulo $\{I\}$. Suppose that $(M, \mathcal{E})$ and $(\bar{M}, \bar{\mathcal{E}})$ are two Monge-Ampere systems, with

$$\mathcal{E} = \{\theta, \Theta, \Psi\}, \quad \bar{\mathcal{E}} = \{\bar{\theta}, \bar{\Theta}, \bar{\Psi}\}.$$

A *Bäcklund transformation* between $(M, \mathcal{E})$ and $(\bar{M}, \bar{\mathcal{E}})$, is a 6-dimensional submanifold $B \subset M \times \bar{M}$ such that in the diagram

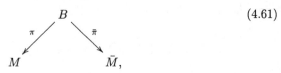

$$\tag{4.61}$$

- each projection $B \to M$, $B \to \bar{M}$ is a submersion; and

- pulled back to B, we have

$$\{\Psi, \bar{\Psi}\} \equiv \{\Theta, \bar{\Theta}\} \pmod{\{\theta, \bar{\theta}\}}.$$

The second condition implies that the dimension of the space of 2-forms spanned by $\{\Theta, \Psi, \bar{\Theta}, \bar{\Psi}\}$ modulo $\{\theta, \bar{\theta}\}$ is at most 2. Therefore,

$$\{\Theta, \Psi\} \equiv \{\bar{\Theta}, \bar{\Psi}\} \pmod{\{\theta, \bar{\theta}\}}.$$

This consequence is what we really want, but the original formulation has the extra benefit of ruling out linear dependence between Θ and $\bar{\Theta}$, which would lead to a triviality in what follows.

A Bäcklund transformation allows one to find a family of integral manifolds of $(\bar{M}, \bar{\mathcal{E}})$ from one integral manifold $N^2 \hookrightarrow M$ of $(M, \mathcal{E})$, as follows. On the 3-dimensional preimage $\pi^{-1}(N) \subset B$, the restriction $\bar{\pi}^*\bar{\mathcal{E}}$ is algebraically generated by $\bar{\theta}$ alone, and is therefore an integrable Pfaffian system. Its integral manifolds can therefore be found by ODE methods, and they foliate $\pi^{-1}(N)$ into a 1-parameter family of surfaces which project by $\bar{\pi}$ to integral manifolds of $(\bar{M}, \bar{\mathcal{E}})$. In each of the following two examples, $(M, \mathcal{E})$ and $(\bar{M}, \bar{\mathcal{E}})$ are equal, so one can generate from one known solution many others.

Example 1: Bäcklund transformation for the sine-Gordon equation.

The primary example concerns the sine-Gordon equation (4.60). The well-known coordinate phenomenon is that if two functions $u(x, y)$, $\bar{u}(x, y)$ satisfy the first-order PDE system

$$\begin{cases} u_x - \bar{u}_x = \lambda \sin(u + \bar{u}), \\ u_y + \bar{u}_y = \frac{1}{\lambda} \sin(u - \bar{u}), \end{cases} \tag{4.62}$$

where $\lambda \neq 0$ is any constant,[11] then each of $u(x, y)$ and $\bar{u}(x, y)$ satisfies (4.60). Conversely, given a function $\bar{u}(x, y)$, the overdetermined system (4.62) for unknown $u(x, y)$ is compatible, and can therefore be reduced to an ODE system, if and only if $\bar{u}(x, y)$ satisfies (4.60). This indicates that given one solution of the sine-Gordon equation, ODE methods give a family of additional solutions.

We fit this example into our definition of a Bäcklund transformation as follows. Start with two copies of the sine-Gordon Monge-Ampère system, one on $M = \{(x, y, u, p, q)\}$ generated by

$$\mathcal{E} = \left\{ \begin{array}{l} \theta = du - p\, dx - q\, dy, \\ \Theta = d\theta = -dp \wedge dx - dq \wedge dy, \\ \Psi = dx \wedge dp + dq \wedge dy - \sin(2u)dx \wedge dy \end{array} \right\},$$

---

[11] This λ will *not* correspond to the integration parameter in the Bäcklund transformation. It plays a role only in the relation to the $K = -1$ system, to be discussed shortly.

the other on $\bar{M} = \{(\bar{x}, \bar{y}, \bar{u}, \bar{p}, \bar{q})\}$ generated by

$$\bar{\mathcal{E}} = \left\{ \begin{array}{l} \bar{\theta} = d\bar{u} - \bar{p}\,d\bar{x} - \bar{q}\,d\bar{y}, \\ \bar{\Theta} = d\bar{\theta} = -d\bar{p} \wedge d\bar{x} - d\bar{q} \wedge d\bar{y}, \\ \bar{\Psi} = d\bar{x} \wedge d\bar{p} + d\bar{q} \wedge d\bar{y} - \sin(2\bar{u})d\bar{x} \wedge d\bar{y} \end{array} \right\}.$$

One can verify that the submanifold $B \subset M \times \bar{M}$ defined by

$$\left\{ \begin{array}{l} \bar{x} = x, \quad \bar{y} = y, \\ p - \bar{p} = \lambda \sin(u + \bar{u}), \\ q + \bar{q} = \frac{1}{\lambda} \sin(u - \bar{u}), \end{array} \right.$$

satisfies the criteria for a Bäcklund transformation, and that the process of solving the overdetermined system (4.62) for $u(x, y)$ corresponds to integrating the Frobenius system as described previously.

For example, the solution $\bar{u}(x, y) = 0$ of sine-Gordon corresponds to the integral manifold $N = \{(x, y, 0, 0, 0)\} \subset M$, whose preimage in $B \subset M \times \bar{M}$ has coordinates (x, y, u) and satisfies

$$\bar{u} = \bar{p} = \bar{q} = 0, \quad p = \lambda \sin(u), \quad q = \frac{1}{\lambda} \sin(u).$$

The system $\mathcal{E}$ is algebraically generated by the form

$$\theta|_{\bar{\pi}^{-1}(N)} = du - \lambda \sin(u)dx - \frac{1}{\lambda} \sin(u)dy.$$

The problem of finding $u(x, y)$ on which this θ vanishes is the same as solving the overdetermined system (4.62) with $\bar{u} = 0$. It is obtained by integrating

$$\frac{du}{\sin u} - \lambda\,dx - \frac{1}{\lambda}\,dy = 0,$$

which has the implicit solution

$$-\ln(\csc u + \cot u) - \lambda x - \frac{1}{\lambda}y = c,$$

where c is the integration constant. This can be solved for u to obtain

$$u(x, y) = 2\tan^{-1}(e^{\lambda x + \frac{1}{\lambda}y + c}).$$

One can verify that this is indeed a solution to the sine-Gordon equation. In principle, we could rename this as $\bar{u}$, and repeat the process to obtain more solutions.

Example 2: Bäcklund transformation for the $K = -1$ system.

Suppose that $f, \bar{f} : N \hookrightarrow \mathbf{E}^3$ are two immersions of a surface into Euclidean space. We say that there is a *pseudospherical line congruence* between $f, \bar{f}$ if for each $p \in N$:

1. the line through $f(p)$ and $\bar{f}(p)$ in $\mathbf{E}^3$ is tangent to each surface at these points (we assume $f(p) \neq \bar{f}(p)$);

2. the distance $r = ||f(p) - \bar{f}(p)||$ is constant;

3. the angle τ between the normals $\nu(p)$ and $\bar{\nu}(p)$ is constant.

This relationship between $f, \bar{f}$ will play a role analogous to that of the system (4.62). We prove the following theorem of Bianchi.

Theorem 4.3 *If there is a pseudospherical line congruence between $f, \bar{f} : N \hookrightarrow$ $\mathbf{E}^3$, then each of f and $\bar{f}$ has constant negative Gauss curvature*

$$K = -\frac{\sin^2(\tau)}{r^2}.$$

It is also true that given one surface $\bar{f}$, there locally exists a surface f sharing a pseudospherical line congruence with $\bar{f}$ if and only if $\bar{f}$ has constant negative Gauss curvature. We will partly verify this claim, after proving Bianchi's theorem.

Proof. Choose Euclidean frame fields $F, \bar{F} : N \to \mathcal{F}$ which are adapted to the pair of surfaces in the sense that

$$\bar{e}_1 = e_1, \tag{4.63}$$

made possible by condition 1 above. Also, as usual, we let $e_3, \bar{e}_3$ be unit normals to $f, \bar{f}$, respectively, which must then satisfy

$$\begin{cases} \bar{e}_2 = (\cos \tau)e_2 + (\sin \tau)e_3, \\ \bar{e}_3 = (-\sin \tau)e_2 + (\cos \tau)e_3, \end{cases} \tag{4.64}$$

with τ constant by condition 3. Now condition 2 says that

$$\bar{f}(p) = f(p) + re_1(p) \tag{4.65}$$

for fixed r. We can use the structure equations

$$df = e_i \cdot \omega^i, \qquad de_i = e_j \cdot \omega_i^j,$$

and similar for $d\bar{f}, d\bar{e}_i$, to obtain relations among the pullbacks by F and $\bar{F}$ of the canonical forms on $\mathcal{F}$. Namely,

$$d\bar{f} = e_1\bar{\omega}^1 + ((\cos \tau)e_2 + (\sin \tau)e_3)\bar{\omega}^2,$$

and also

$$d\bar{f} = d(f + re_1) = e_1\omega^1 + e_2(\omega^2 + r\omega_1^2) + e_3(r\omega_1^3),$$

so that

$$\begin{cases} \bar{\omega}^1 = \omega^1, \\ (\cos \tau)\bar{\omega}^2 = \omega^2 + r\omega_1^2, \\ (\sin \tau)\bar{\omega}^2 = r\omega_1^3. \end{cases} \tag{4.66}$$

Note that $\omega^2 + r\omega_1^2 = (r \cot \tau)\omega_1^3$, and this gives a necessary condition on f alone to share a pseudospherical line congruence. Similar calculations using $e_1 = \bar{e}_1$ yield

$$\begin{cases} \bar{\omega}_1^2 = (\cos \tau)\omega_1^2 + (\sin \tau)\omega_1^3, \\ \bar{\omega}_1^3 = -(\sin \tau)\omega_1^2 + (\cos \tau)\omega_1^3, \end{cases}$$

and differentiating the remaining relations (4.64) gives

$$\bar{\omega}_2^3 = \omega_2^3,$$

giving complete expressions for $\bar{F}^*\omega$ in terms of $F^*\omega$. Note in particular that

$$\bar{\omega}_1^3 = \frac{\sin \tau}{r}\omega^2.$$

Now we can consider the curvature, expanding both sides of the definition

$$d\bar{\omega}_1^2 = -\bar{K}\bar{\omega}^1 \wedge \bar{\omega}^2. \tag{4.67}$$

First,

$$\begin{aligned} d\bar{\omega}_1^2 &= -\bar{\omega}_3^2 \wedge \bar{\omega}_1^3 \\ &= -\omega_3^2 \wedge \frac{\sin \tau}{r}\omega^2 \\ &= h_{21}\frac{\sin \tau}{r}\omega^1 \wedge \omega^2, \end{aligned}$$

where h_{21} is part of the second fundamental form of $F : N \hookrightarrow M$, defined by $\omega_i^3 = h_{ij}\omega^j$. Note in particular that if $h_{12} = 0$, then ω_1^3 is a multiple of ω^1, so by (4.66), $\bar{\omega}^1 \wedge \bar{\omega}^2 = 0$, a contradiction; we can now assume that $h_{12} = h_{21} \neq 0$. On the right-hand side of (4.67),

$$\begin{aligned} -\bar{K}\bar{\omega}^1 \wedge \bar{\omega}^2 &= -\bar{K}\omega^1 \wedge \frac{r}{\sin \tau}\omega_1^3 \\ &= -\bar{K}h_{12}\frac{r}{\sin \tau}\omega^1 \wedge \omega^2. \end{aligned}$$

Equating these expressions, we have

$$\bar{K} = -\frac{\sin^2 \tau}{r^2}$$

as claimed.   □

Now suppose given a surface $f : N \hookrightarrow \mathbf{E}^3$ with constant negative Gauss curvature $K = -1$. We are interested in finding $\bar{f}$ which shares with f a pseudospherical line congruence.

We start with local coordinates (s, t) on N whose coordinate lines $ds = 0$, $dt = 0$ define the asymptotic curves of f. It will be convenient to instead have orthogonal coordinate lines, so we define the coordinates $x = s + t$, $y = s - t$, for which $dx = 0$ and $dy = 0$ define the principal curves of f. We have seen in (4.55, 4.56) that the first and second fundamental forms are given by

$$
\begin{aligned}
I &= ds^2 + 2\cos(2z)ds\,dt + dt^2 & (4.68)\\
&= \cos^2 z\,dx^2 + \sin^2 z\,dy^2, & (4.69)\\
II &= 2\sin(2z)ds\,dt & (4.70)\\
&= \sin(z)\cos(z)(dx^2 - dy^2), & (4.71)
\end{aligned}
$$

where $2z$ is half of the angle measure between the asymptotic directions and satisfies the sine-Gordon equation. One orthonormal coframing is given by $(\cos(z)dx, \sin(z)dy)$; we consider an orthonormal coframing differing from this one by rotation by some α:

$$
\begin{aligned}
\begin{pmatrix} \omega^1 \\ \omega^2 \end{pmatrix} &= \begin{pmatrix} \cos\alpha & \sin\alpha \\ -\sin\alpha & \cos\alpha \end{pmatrix} \begin{pmatrix} \cos(z)dx \\ \sin(z)dy \end{pmatrix} & (4.72)\\
&= \begin{pmatrix} \cos(\alpha - z) & \cos(\alpha + z) \\ -\sin(\alpha - z) & -\sin(\alpha + z) \end{pmatrix} \begin{pmatrix} ds \\ dt \end{pmatrix}. & (4.73)
\end{aligned}
$$

The idea here is that we are looking for a function α on N for which this coframing could be part of that induced by a pseudospherical line congruence. The main compatibility condition, derived from (4.66), is

$$
\omega^2 + r\omega_1^2 = r(\cot \tau)\omega_1^3. \tag{4.74}
$$

We can compute the Levi-Civita connection form ω_1^2 using the structure equations $d\omega^i = -\omega_j^i \wedge \omega^j$, and find

$$
\omega_1^2 = (\alpha_s + z_s)ds + (\alpha_t - z_t)dt.
$$

Similarly, we can compute from (4.70) and (4.73) the coefficients of the second fundamental form with respect to the coframe (ω^1, ω^2), and find

$$
\begin{aligned}
\omega_1^3 &= \sin(\alpha - z)ds - \sin(\alpha + z)dt,\\
\omega_2^3 &= \cos(\alpha - z)ds - \cos(\alpha + z)dt.
\end{aligned}
$$

Substituting these into the compatibility condition (4.74), we obtain an equation of 1-forms whose ds, dt coefficients are

$$
\begin{aligned}
\alpha_s + z_s &= (\csc\tau + \cot\tau)\sin(\alpha - z),\\
\alpha_t - z_t &= \left(\frac{1}{\csc\tau + \cot\tau}\right)\sin(\alpha + z).
\end{aligned}
$$

We compare this to (4.62), and conclude that the local existence of a solution α is equivalent to having z satisfy the sine-Gordon equation. Note that the role played by λ in (4.62) is similar to that played by the angle τ in the pseudospherical line congruence. We conclude by exhibiting a surprising use of the Bäcklund transformation for the $K = -1$ system. This starts with an integral

manifold in M^5 of $\mathcal{E}$ (see (4.49)) that is *not* transverse as a Legendre submanifold, in the sense of being a 1-jet lift of an immersed surface in $\mathbf{E}^3$. Instead,
regarding the contact manifold M as the unit sphere bundle over $\mathbf{E}^3$, $N \hookrightarrow M$
consists of the unit normal bundle of the line $\{(0, 0, w) : w \in \mathbf{R}\} \subset \mathbf{E}^3$. This
Legendre surface is topologically a cylinder. To study its geometry, we will work
in the circle bundle $\mathcal{F} \to M$. Its preimage there is parameterized by

$$(u, v, w) \mapsto \left((0, 0, w), \begin{pmatrix} e_1 = (\sin u \cos v, -\sin u \sin v, \cos u), \\ e_2 = (\cos u \cos v, -\cos u \sin v, -\sin u), \\ e_3 = (\sin v, \cos v, 0) \end{pmatrix} \right), \quad (4.75)$$

where $(u, v, w) \in S^1 \times S^1 \times \mathbf{R}$. It is easily verified that this is an integral manifold
for the pullback of $\mathcal{E}$ by $\mathcal{F} \to M$. We will apply the Bäcklund transformation
to this degenerate integral manifold, and obtain a non-trivial surface in $\mathbf{E}^3$ with
Gauss curvature $K = -1$.

We will take the Bäcklund transformation to be the submanifold $B \subset \mathcal{F} \times \bar{\mathcal{F}}$
($\bar{\mathcal{F}}$ is another copy of $\mathcal{F}$) defined by (4.63, 4.64, 4.65); this is a lift of the original
picture (4.61) from M to $\mathcal{F}$. We fix the constants of the line congruence to be

$$\tau = \tfrac{\pi}{2}, \quad r = 1.$$

As a consequence, if our Bäcklund transformation gives a transverse Legendre
submanifold, then Theorem 4.3 states that the corresponding surface in $\mathbf{E}^3$ will
have Gauss curvature $K = -1$.

Now, the definition of $B \subset \mathcal{F} \times \bar{\mathcal{F}}$ provides a unique lift $\pi^{-1}(N)$ of our
degenerate integral manifold (4.75) to B. The "other" $K = -1$ system $\bar{\mathcal{E}}$ should
restrict to $\pi^{-1}(N) \subset B \subset \mathcal{F} \times \bar{\mathcal{F}}$ to be algebraically generated by the 1-form
$\bar{\theta}$, and then $\pi^{-1}(N)$ will be foliated into surfaces which project into integral
manifolds of $\bar{\mathcal{E}}$. So we compute $\bar{\theta} = \bar{\omega}^3$:

$$\begin{aligned} \bar{\omega}^3 &= \langle d\bar{x}, \bar{e}_3 \rangle \\ &= \langle dx + de_1, -e_2 \rangle \\ &= \sin u \, dw - du. \end{aligned}$$

Indeed, this 1-form is integrable, and its integral manifolds are of the form

$$u = 2 \tan^{-1}(\exp(w + c)), \quad (4.76)$$

where c is an integration constant. We will consider the integral manifold corresponding to $c = 0$. The Euclidean surface that we are trying to construct is
now parameterized by $\bar{x}(u, v, w)$, constrained by (4.76). We obtain

$$\begin{aligned} \bar{x}(u, v, w) &= x(u, v, w) + e_1(u, v, w) \\ &= (0, 0, w) + (\sin u \cos v, -\sin u \sin v, \cos u) \\ &= \left(\tfrac{2e^w}{1 + e^{2w}} \cos v, -\tfrac{2e^w}{1 + e^{2w}} \sin v, w + \tfrac{1 - e^{2w}}{1 + e^{2w}} \right) \\ &= (\operatorname{sech} w \cos v, -\operatorname{sech} w \sin v, w - \tanh w). \end{aligned}$$

This surface in $\mathbf{E}^3$ is the pseudosphere, the most familiar surface of constant negative Gauss curvature; we introduced both it and the "framed line" in §1.4, as examples of smooth but non-transverse Legendre submanifolds of the unit sphere bundle $M \to \mathbf{E}^3$. In principle, we could iterate this Bäcklund transformation, obtaining arbitrarily many examples of $K = -1$ surfaces.

Bibliography

[AI79]   Robert L. Anderson and Nail H. Ibragimov, *Lie-bäcklund transforma-tions in applications*, Studies in Applied Mathematics, no. 1, SIAM, Philadelphia, 1979. MR 80e:58048 196

[And]   Ian M. Anderson, *The variational bicomplex*, Preprint available at http://www.math.usu.edu. 6

[And92]   Ian M. Anderson, *Introduction to the variational bicomplex*, Mathe-matical aspects of classical field theory (Seattle, WA, 1991), Amer. Math. Soc., Providence, RI, 1992, pp. 51–73. MR 94a:58045 6

[B⁺91]   Robert L. Bryant et al., *Exterior differential systems*, MSRI Publica-tions, no. 18, Springer-Verlag, New York, 1991. MR 92h:58007 10, 162, 179

[Bet84]   David E. Betounes, *Extension of the classical Cartan form*, Phys. Rev. D (3) **29** (1984), 599–606. MR 86g:58043 169

[Bet87]   _____, *Differential geometric aspects of the Cartan form: Symmetry theory*, J. Math. Phys. **28** (1987), 2347–2353. MR 88k:58036 169

[BG95a]   Robert L. Bryant and Phillip A. Griffiths, *Characteristic cohomol-ogy of differential systems, I: General theory*, J. Amer. Math. Soc. **8** (1995), 507–596. MR 96c:58183 184

[BG95b]   _____, *Characteristic cohomology of differential systems, II: Con-servation laws for a class of parabolic equations*, Duke Math. J. **78** (1995), 531–676. MR 96d:58158 194

[BGH95]   Robert L. Bryant, Phillip A. Griffiths, and Lucas Hsu, *Hyperbolic exterior differential systems and their conservation laws, I,II*, Selecta Math. (N.S.) **1** (1995), 21–112, 265–323. MR 97d:580008,97d:580009 191

[Bla67]   Wilhelm Blaschke, *Vorlesungen über Differentialgeometrie und ge-ometrische Grundlagen von Einsteins Relativitätstheorie*, Chelsea Publishing Co., New York, 1967. 69

207

[Car33] Élie Cartan, *Les espaces métriques fondés sur la notion d'aire*, Exposés de Géométrie, I, Hermann, Paris, 1933. 5

[Car34] ———, *Les espaces de Finsler*, Exposés de Géométrie, II, Hermann, Paris, 1934. 5

[Car71] ———, *Leçons sur les invariants intégraux*, Hermann, Paris, 1971, troisième tirage. MR 50 #8238 5

[Chr86] Demetrios Christodoulou, *Global solutions of nonlinear hyperbolic equations for small initial data*, Comm. Pure Appl. Math. **39** (1986), 267–282. MR 87c:35111 126

[Ded77] Paul Dedecker, *On the generalization of symplectic geometry to multiple integrals in the calculus of variations*, Differential geometrical methods in mathematical physics (Proc. Sympos., Univ. Bonn, Bonn, 1975), Springer, Berlin, 1977, pp. 395–456. Lecture Notes in Math., Vol. 570. MR 56 #16680 6

[Don35] Théophile de Donder, *Théorie invariantive du calcul des variations*, Gauthier-Villars, Paris, 1935, Deuxième édition. 6

[GH96] Mariano Giaquinta and Stefan Hildebrandt, *Calculus of variations, I,II*, Grundlehren der Mathematischen Wissenschaften, no. 310, 311, Springer-Verlag, Berlin, 1996. MR 98b:49002a,b 153, 158

[Gra00] Michele Grassi, *Local vanishing of characteristic cohomology*, Duke Math. J. **102** (2000), 307–328. MR 2001j:58031 163, 171

[Gri83] Phillip A. Griffiths, *Exterior differential systems and the calculus of variations*, Progress in Mathematics, vol. 25, Birkhäuser, Boston, 1983. MR 84h:58007 7

[Joh79] Fritz John, *Blow-up of solutions of nonlinear wave equations in three space dimensions*, Manuscripta Math. **28** (1979), 235–268. MR 80i:35114 137

[Lep46] Théophile Lepage, *Sur une classe d'équations aux dérivées partielles du second ordre*, Acad. Roy. Belgique, Bull. Cl. Sci. (5) **32** (1946), 140–151. 6

[Lep54] ———, *Équation du second ordre et tranformations symplectiques*, Premier colloque sur les équations aux dérivées partielles, Louvain, 1953, Masson & Cie, Paris, 1954, pp. 79–104. MR 16,1028a 6

[Lev74] Howard A. Levine, *Instability and nonexistence of global solutions to nonlinear wave equations of the form $Pu_{tt} = -Au + \mathcal{F}(u)$*, Trans. Amer. Math. Soc. **192** (1974), 1–21. MR 49 #9436 136

[LRC93] V. V. Lychagin, V. N. Rubtsov, and I. V. Chekalov, *A classification of Monge-Ampère equations*, Ann. Sci. École Norm. Sup. (4) **26** (1993), 281–308. MR 94c:58229 45

[Olv93] Peter J. Olver, *Applications of lie groups to differential equations*, Graduate Texts in Mathematics, no. 107, Springer-Verlag, New York, 1993, second edition. MR 94g:58260 176, 185

[Poh65] Stanislav I. Pohožaev, *On the eigenfunctions of the equation* $\Delta u + \lambda f(u) = 0$, Soviet Math. Dokl. **6** (1965), 1408–1411. MR 33 #411 125

[Rum90] Michel Rumin, *Un complexe de formes différentielles sur les variétés de contact*, C. R. Acad. Sci. Paris Sér. I Math. **310** (1990), 401–404. MR 91a:58004 17

[Spi75] Michael Spivak, *A comprehensive introduction to differential geometry*, vol. IV, Publish or Perish, Inc., Boston, 1975. MR 52 #15254a 148

[Str89] Walter A. Strauss, *Nonlinear wave equations*, CBMS Regional Conference Series in Mathematics, no. 73, American Mathematical Society, Providence, RI, 1989. MR 91g:35002 129

[Via00] Jeff A. Viaclovsky, *Conformal geometry, contact geometry, and the calculus of variations*, Duke Math. J. **101** (2000), no. 2, 283–316. MR 2001b:53038 107

[Vin84] Alexandre M. Vinogradov, *The C-spectral sequence, Lagrangian formalism, and conservation laws, I,II*, J. Math. Anal. Appl. **100** (1984), 1–40, 41–129. MR 85j:58150a,b 176, 183, 185

[Woo94] John C. Wood, *Harmonic maps into symmetric spaces and integrable systems*, Harmonic Maps and Integrable Systems (Allan P. Fordy and John C. Wood, eds.), Aspects of Mathematics, vol. 23, Vieweg, Braunschweig/Weisbaden, 1994, pp. 29–55. MR 95m:58047 (collection) 176

Index